AF361776

Intelligent Autonomous Decision-Making and Cooperative Control Technology of High-Speed Vehicle Swarms

Intelligent Autonomous Decision-Making and Cooperative Control Technology of High-Speed Vehicle Swarms

Editors

Dong Zhang
Wei Huang

MDPI • Basel • Beijing • Wuhan • Barcelona • Belgrade • Manchester • Tokyo • Cluj • Tianjin

Editors
Dong Zhang
Northwestern Polytechnical University
China

Wei Huang
National University of Defense Technology
China

Editorial Office
MDPI
St. Alban-Anlage 66
4052 Basel, Switzerland

This is a reprint of articles from the Special Issue published online in the open access journal *Applied Sciences* (ISSN 2076-3417) (available at: https://www.mdpi.com/journal/applsci/special_issues/Cooperative_Control_Technology_High_Speed_Vehicle_Swarms).

For citation purposes, cite each article independently as indicated on the article page online and as indicated below:

LastName, A.A.; LastName, B.B.; LastName, C.C. Article Title. *Journal Name* **Year**, *Volume Number*, Page Range.

ISBN 978-3-0365-4231-7 (Hbk)
ISBN 978-3-0365-4232-4 (PDF)

Contents

About the Editors

Dong Zhang

He is currently an associate professor and doctoral supervisor of Northwestern Polytechnical University. He is also a visiting professor of the Key Laboratory of Complex System Control and Intelligent Collaborative Technology. His current research interests include the dynamic modeling and control of aerospace vehicles, high-speed aircraft swarms combat mission planning, and collaborative guidance and control. He has presided over more than 10 national projects, published more than 30 academic papers and 1 academic monograph, and won the second prize for military-scientific technological progress in 2020.

Wei Huang

He is currently a professor at the National University of Defense Technology. His current research interests include wind turbines, vortex, hypersonics, drag, vorticity, numerical simulation, flow, aerodynamics, aircraft, and drag reduction.

applied sciences

Editorial

Special Issue on "Intelligent Autonomous Decision-Making and Cooperative Control Technology of High-Speed Vehicle Swarms"

Dong Zhang [1,*] and Wei Huang [2]

1 School of Astronautics, Northwestern Polytechnical University, Xi'an 710072, China
2 College of Aerospace Science and Engineering, National University of Defense Technology, Xi'an 710072, China; gladrain2001@163.com
* Correspondence: zhangdong@nwpu.edu.cn; Tel.: +86-29-88492788

Citation: Zhang, D.; Huang, W. Special Issue on "Intelligent Autonomous Decision-Making and Cooperative Control Technology of High-Speed Vehicle Swarms". *Appl. Sci.* **2022**, *12*, 4409. https://doi.org/10.3390/app12094409

Received: 14 April 2022
Accepted: 15 April 2022
Published: 27 April 2022

Publisher's Note: MDPI stays neutral with regard to jurisdictional claims in published maps and institutional affiliations.

Swarm intelligence technology is a high and new technology combining unmanned system technology, network information technology and artificial intelligence technology, which has become a research hotspot. The characteristics of high-speed and high dynamic flight, large airspace flight environment and multi-field coupling battlefields under strong confrontation bring great challenges to the cooperative operation of high-speed vehicle swarms.

Due to the high speed and high dynamic flight characteristics, as well as the large airspace flight environment, the fluid–thermal–structural coupling effect makes it difficult for the traditional UAV swarm technology to be directly applied to the high-speed aircraft swarm system, which brings great challenges to its decision making and planning. Therefore, it is urgent to study new theories and methods for the cooperative decision making and mission planning of high-speed aircraft swarm systems.

This Special Issue, "Intelligent autonomous decision-making and cooperative control technology of high-speed vehicle swarms", has been launched to provide an opportunity for researchers in the area of collaborative decision-making and mission-planning technology for high-speed aircraft swarms to highlight recent developments made in their fields. Eight excellent papers that cover a wide variety of characteristic analysis and parameter optimization method for hypersonic vehicle/fully distributed control/multi-constraints cooperative guidance method/formation control technology aspects were selected for publication in this Special Issue [1–7]. These eights papers have been summarized as follows:

- Li et al. [1] proposes a parameter-optimization method for air-breathing hypersonic vehicle. Their method uses a neural network to model the relationship between the aircraft parameters and optimal cruise point and can provide good guidance for the adjustment of aircraft parameters.
- Cong et al. [2] studies a multi-constraints cooperative guidance method based on distributed MPC for multi-missiles. The method can simultaneously control multiple missiles to perform attacks on the target with the constraints of impact time and impact angle on the premise of meeting the requirements of miss distance.
- Liu et al. [3] proposes a novel distributed consensus method for MAS with completely unknown system nonlinearities and time-varying control coefficients under a directed graph. It is rigorously proved that the consensus of the MAS is achieved while guaranteeing the prescribed tracking-error performance.
- Suo et al. [4] proposes a route-based formation switching and obstacle avoidance method for the formation control problem of fixed-wing UAVs in distributed ad hoc networks. The results shows that the method is helpful to deal with the communication anomalies and flight anomalies with variable topology.
- Qin et al. [5] proposes a distributed grouping cooperative dynamic task assignment method based on extended contract network protocol. The method can perform reconnaissance-and-attack tasks to multi-targets in complex and uncertain combat

scenarios, improve the adaptiveness of the swarm under the sudden circumstance, and realize the optimization of the task execution efficiency of the UAV swarm.

- Luo et al. [6] proposes a UAV-cooperative penetration dynamic-tracking interceptor method based on DDPG, which can realize the time coordination of multi-UAV cooperative penetration.
- Zhang et al. [7] investigates the air–ground cooperative time-varying formation-tracking control problem of a heterogeneous cluster system composed of a UGV and a UAV. Using a linear quadratic optimal control theory, a UAV–UGV formation–maintenance controller is designed to track the reference trajectory of the UGV based on the UAV–UGV relative motion model.

Funding: This research received no external funding.

Conflicts of Interest: The authors declare no conflict of interest.

References

1. Li, H.; Zhou, Y.; Wang, Y.; Du, S.; Xu, S. Optimal Cruise Characteristic Analysis and Parameter Optimization Method for Air-Breathing Hypersonic Vehicle. *Appl. Sci.* **2021**, *11*, 9565. [CrossRef]
2. Cong, M.; Cheng, X.; Zhao, Z.; Li, Z. Studies on Multi-Constraints Cooperative Guidance Method Based on Distributed MPC for Multi-Missiles. *Appl. Sci.* **2021**, *11*, 10857. [CrossRef]
3. Liu, Z.; Huang, H.; Luo, S.; Fu, W.; Li, Q. Fully Distributed Control for a Class of Uncertain Multi-Agent Systems with a Directed Topology and Unknown State-Dependent Control Coefficients. *Appl. Sci.* **2021**, *11*, 11304. [CrossRef]
4. Suo, W.; Wang, M.; Zhang, D.; Qu, Z.; Yu, L. Formation Control Technology of Fixed-Wing UAV Swarm Based on Distributed Ad Hoc Network. *Appl. Sci.* **2022**, *12*, 535. [CrossRef]
5. Qin, B.; Zhang, D.; Tang, S.; Wang, M. Distributed Grouping Cooperative Dynamic Task Assignment Method of UAV Swarm. *Appl. Sci.* **2022**, *12*, 2865. [CrossRef]
6. Luo, Y.; Song, J.; Zhao, K.; Liu, Y. UAV-Cooperative Penetration Dynamic-Tracking Interceptor Method Based on DDPG. *Appl. Sci.* **2022**, *12*, 1618. [CrossRef]
7. Zhang, J.; Yue, X.; Zhang, H.; Xiao, T. Optimal Unmanned Ground Vehicle-Unmanned Aerial Vehicle Formation-Maintenance Control for Air-Ground Cooperation. *Appl. Sci.* **2022**, *12*, 3598. [CrossRef]

Article

Studies on Multi-Constraints Cooperative Guidance Method Based on Distributed MPC for Multi-Missiles

Mingyu Cong [1,2,3], Xianghong Cheng [1,2,*], Zhiquan Zhao [3] and Zhijun Li [3]

1   School of Instrument Science & Engineering, Southeast University, Nanjing 210096, China; 230188067@seu.edu.cn
2   Key Laboratory of Micro-Inertial Instrument and Advanced Navigation Technology, Ministry of Education, Southeast University, Nanjing 210096, China
3   Jiangxi Hongdu Aviation Industry Group, Nanchang 330024, China; airlqbz@163.com (Z.Z.); lizhijunde@163.com (Z.L.)
*   Correspondence: xhcheng@seu.edu.cn

Abstract: Cooperative terminal guidance with impact angle constraint is a key technology to achieve a saturation attack and improve combat effectiveness. The present study envisaged cooperative terminal guidance with impact angle constraint for multiple missiles. In this pursuit, initially, the three-dimensional cooperative terminal guidance law with multiple constraints was studied. The impact time cooperative strategy of virtual leader missile and follower missiles was designed by introducing virtual leader missiles. Subsequently, based on the distributed model prediction control combined with the particle swarm optimization algorithm, a cooperative terminal guidance algorithm was designed for multiple missiles with impact angle constraint that met the guidance accuracy. Finally, the effectiveness of the algorithm was verified using simulation experiments.

Keywords: cooperative guidance; model prediction control; multi-missile cooperative control; multi-constraint cooperative guidance

Citation: Cong, M.; Cheng, X.; Zhao, Z.; Li, Z. Studies on Multi-Constraints Cooperative Guidance Method Based on Distributed MPC for Multi-Missiles. *Appl. Sci.* **2021**, *11*, 10857. https://doi.org/10.3390/app112210857

Academic Editors: Wei Huang and Dong Zhang

Received: 18 October 2021
Accepted: 8 November 2021
Published: 17 November 2021

Publisher's Note: MDPI stays neutral with regard to jurisdictional claims in published maps and institutional affiliations.

1. Introduction

Cooperative guidance is one of the key technologies for multiple missiles to attack and intercept large maneuvering targets with a high accuracy [1,2]. The two major types of cooperative guidance technology that have been extensively studied include independent cooperative guidance and distributed cooperative guidance.

Independent cooperative guidance determines the main guidance information by relying on only its own individual information, and there is no interaction of information between the missiles. The attack tasks are completed independently according to the designated guidance law and impact time before the launch. Wu S.T. et al. [3] was the first to propose the guidance problem along with the impact time. After more than ten years of development, many research results have emerged in this field. Based on the combination of the optimal control and time adjustment, Ma G.X. et al. [4] designed a cooperative guidance law with the impact time and angle constraints, and realized the cooperative interception based on the impact angle constraints. Li G.Y. et al. [5] designed the cooperative guidance law for multiple missiles based on the terminal sliding mode control method and the principle of consistency. They obtained the line-of-sight (LOS) angle rate for all the cooperative missiles converging to zero in a finite time, and it was found that the LOS angle converged to the desired angle. Zhou J. et al. [6] proposed a cooperative guidance law based on "leader-follower" architecture and sliding mode control to achieve a cooperative attack with line-of-sight (LOS) angle constraints on moving targets. Based on the impact angle, a guidance law with the impact time based on the sliding mode control was designed [7,8]. During the flight, multiple missiles exchange information with each other, "negotiate" together, and adjust their impact time according to a certain strategy in order to achieve simultaneous arrival and saturation attack. This is called distributed cooperative guidance. Various studies that have been conducted on distributed cooperative guidance includes the time cooperative guidance law based on bias proportional guidance, the time cooperative guidance law based on the leader-follower and the time based on switching logic. Using a weighted average consensus algorithm, a distributed time cooperative guidance law was designed [9,10]. Wang X. et al. [11] designed the guidance law with the impact time based on a bias proportional guidance. Assuming that each missile had the same speed, Tekin R. et al. [12] designed a time cooperative guidance law based on the 'leader-follower' model. Jeon J.S. et al. [13] considered the integration of the time cooperative guidance and control and studied the tracking problem of the nominal projectile distance.

The guidance law with impact time constraints is a key technology in achieving saturation attack, break through anti-missile defense systems, and accomplishing reliable strikes against important strategic targets. Lin D.F. et al. [14] designed a cooperative guidance law with an impact time control and field of view constraints of the seeker for the cooperative guidance of multiple missiles in attacking low-speed targets. Li G.Y. et al. [15] designed the cooperative guidance law for multiple missiles to simultaneously intercept high maneuvering targets and designed the cooperative guidance law for multiple missiles based on the terminal sliding mode control method and the principle of consistency to achieve the LOS angle rate of the cooperative missiles converging to zero in a finite time. In view of the Field of View (FOV) constraints of the missiles, the terminal LOS angle constraints [16–18] were employed in the design of the cooperative guidance law, along with the impact time constraints.

Considering the research on the cooperative guidance algorithm, there are more studies on the guidance law with impact angle or time constraints [19–26], and relatively fewer studies on combining the two as terminal constraints [27–31]. Impact time constraints are the basis of the cooperative guidance algorithm and are indispensable. Impact angle constraints can make the missile hit the target with the best attitude in order to maximize the effectiveness of the warhead to achieve a maximum damage. Hence, considering both impact time and impact angle constraints in the cooperative terminal guidance algorithm is of great significance for improving the effective damage and combat capability of the cruise missile weapon system.

Model prediction control (MPC), also known as receding horizon control (RHC), has an online rolling optimization at the core, which is essentially an optimal control problem in the finite time domain. At each optimal control moment, the current optimal control domain can be obtained, and the first part of the domain is used as the optimal control to act on the system. The optimal control sequence is obtained after multiple sampling. MPC is a model-based algorithm that has been widely used in industry, metallurgy, etc. [32]. In recent years, MPC has been continuously improved, and a few improved methods have been for proposed for MPC [33–36], which achieved better application prospects and obtained better prediction results.

In the present study, we proposed a three-dimensional cooperative terminal guidance law that met the constraints of impact angle and impact time simultaneously, on the premise of meeting the requirements of miss distance. First, the nonlinear motion model of the missile and the target was established, the state quantity and the control quantity were normalized, the concept of leader-follower missiles was introduced, and the impact time cooperative strategy of leader and follower missiles was designed. Subsequently, based on the model predictive control method, a three-dimensional cooperative terminal guidance law was proposed that simultaneously controlled multiple missiles to perform attacks on the target with the constraints of impact time and impact angle on the premise of meeting the requirements of miss distance. Finally, by introducing one leader missile and four follower missiles to simulate the multi-missile cooperative attack against stationary targets, linear moving targets, and snake-shaped maneuvering targets, the effectiveness of the multi-constraint cooperative guidance law algorithm was verified.

2. Description of Cooperative 3D Terminal Guidance for Multiple Missiles

The relative motion relationship between the missile and the target is shown in Figure 1. Herein, M and T represent the missile and the target, respectively. The ground inertial coordinate system is expressed in the *A-XYZ* plane. This is the relative distance between the missile and the target, is the line-of-sight elevation angle and the line-of-sight azimuth angle, represents the trajectory inclination angle and the trajectory deflection angle of the missile, and represents the trajectory inclination angle and the trajectory deflection angle of the target. Unless otherwise specified, all angles were positive counterclockwise.

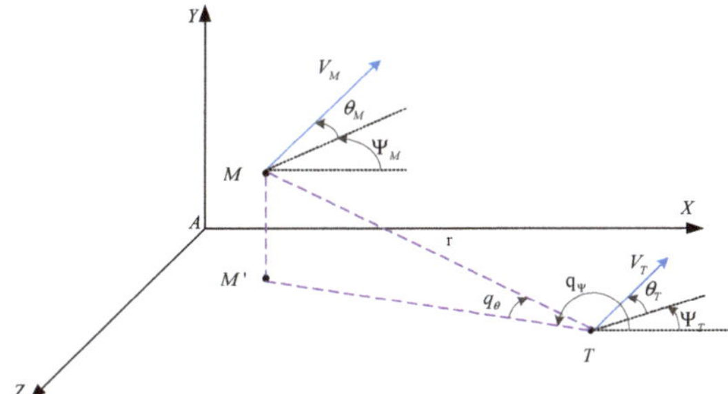

Figure 1. Guidance geometric relationship.

According to the relative geometric relationship between the missile and the target in Figure 1, the relative motion can be obtained as:

$$
\begin{cases}
\frac{dL}{dt} = \frac{d}{dt}(r\boldsymbol{i}_L) = \boldsymbol{V}_T\boldsymbol{i}_T - \boldsymbol{V}_M\boldsymbol{i}_M = \dot{r}\boldsymbol{i}_L + \boldsymbol{\Omega}_L \times r\boldsymbol{i}_L \\
\boldsymbol{A}_T = A_{yt}\boldsymbol{j}_T + A_{zt}\boldsymbol{k}_T = \boldsymbol{\Omega}_L \times \boldsymbol{V}_T + \boldsymbol{\Omega}_T \times \boldsymbol{V}_T \\
\boldsymbol{A}_M = A_{ym}\boldsymbol{j}_M + A_{zm}\boldsymbol{k}_M = \boldsymbol{\Omega}_L \times \boldsymbol{V}_M + \boldsymbol{\Omega}_M \times \boldsymbol{V}_M \\
\boldsymbol{\Omega}_L = \dot{\psi}_L \sin\theta_L \boldsymbol{i}_L - \dot{\theta}_L \boldsymbol{j}_L + \dot{\psi}_L \cos\theta_L \boldsymbol{k}_L = \dot{\lambda}_x \boldsymbol{i}_L + \dot{\lambda}_y \boldsymbol{j}_L + \dot{\lambda}_z \boldsymbol{k}_L \\
\boldsymbol{\Omega}_M = \dot{\psi}_m \sin\theta_m \boldsymbol{i}_M - \dot{\theta}_m \boldsymbol{j}_M + \dot{\psi}_m \cos\theta_m \boldsymbol{k}_M \\
\boldsymbol{\Omega}_T = \dot{\psi}_t \sin\theta_t \boldsymbol{i}_T - \dot{\theta}_t \boldsymbol{j}_T + \dot{\psi}_t \cos\theta_t \boldsymbol{k}_T
\end{cases}
\tag{1}
$$

where $\boldsymbol{V}_T$ and $\boldsymbol{V}_M$ are the velocity vectors of the target and the missile, respectively; A_{zm} and A_{ym} are the normal accelerations of the turning plane and the dive plane of the missile, respectively; A_{zt} and A_{yt} are the normal accelerations of the target turning plane and the dive plane, respectively; $\boldsymbol{\Omega}_L$ is the rotation angular velocity vector of the line-of-sight coordinate system; $\boldsymbol{\Omega}_T$ is the rotation angular velocity vector of the target relative to the line of sight coordinate system; and $\boldsymbol{\Omega}_M$ is the rotation angular velocity vector of the missile relative to the line of sight coordinate system.

From Equation (1), the following differential equations can be obtained:

$$
\begin{cases}
\dot{r} = V_m \cdot (\rho\cos\theta_t\cos\psi_t - \cos\theta_m\cos\psi_m) \\
r\dot{\lambda}_y = V_m \cdot (\sin\theta_m - \rho\sin\theta_t) \\
r\dot{\lambda}_z = V_m \cdot (\rho\cos\theta_t\sin\psi_t - \cos\theta_m\sin\psi_m) \\
\dot{\theta}_m = \frac{A_{zm}}{V_m} + \frac{1}{r}V_m\tan\lambda_y\sin\psi_m(\rho\cos\theta_t\sin\psi_t - \cos\theta_m\sin\psi_m) - \frac{1}{r}V_m\cos\psi_m(\rho\sin\theta_t - \sin\theta_m) \\
\dot{\psi}_m = \frac{A_{ym}}{V_m\cos\theta_m} - \frac{1}{r\cos\theta_m}V_m\sin\theta_m\cos\psi_m \times \tan\lambda_y(\rho\cos\theta_t\sin\psi_t - \cos\theta_m\sin\psi_m)\ldots \\
\quad - \frac{1}{r\cos\theta_m}V_m\sin\theta_m\sin\psi_m(\rho\sin\theta_t - \sin\theta_m) - \frac{1}{r}V_m(\rho\cos\theta_t\sin\psi_t - \cos\theta_m\sin\psi_m) \\
\dot{\theta}_t = \frac{A_{zt}}{V_t} + \frac{1}{r}V_m\sin\psi_t\tan\lambda_y(\rho\cos\theta_t\sin\psi_t - \cos\theta_m\sin\psi_m) - \frac{1}{r}V_m\cos\psi_t(\rho\sin\theta_t - \sin\theta_m)
\end{cases}
\tag{2}
$$

where $\rho = V_t/V_m$ is the speed ratio of the target relative to the missile; $\boldsymbol{A}_M$ and $\boldsymbol{A}_T$ are the acceleration vectors of the missile and the target, respectively; and A_{zmd} and A_{ymd} are the acceleration commands for the longitudinal and lateral channels of the missile, respectively.

3. Strategy for Timed Cooperative Guidance of Multiple Missiles

For cooperative guidance, one of the important features is the ability wherein the multiple missiles can hit the target simultaneously. In order to achieve this goal, the present study employed the concept of virtual leader missile. According to the selection principle, one of the missiles was selected as the virtual leader missile, and the traditional guidance scheme was adopted to make it fly towards the target, while the remaining missiles were regarded as follower missiles. In the dive plane, the follower missiles used the same guidance algorithm as the virtual leader missile. In the turning plane, the follower missiles realized the time cooperation through lateral maneuvering (which was mainly achieved by controlling the relative distance between the missile and the target).

3.1. Movement Law of the Virtual Leader Missile

The relative motion relationship of the leader-follower missiles can be seen form Figure 2. Considering that Missles Leader (ML) is the leader missile, Mi is the ith follower missile, and T is the target, the velocity of the leader missile is V_{m0}, the trajectory inclination angle and the trajectory deflection angle are θ_0 and ψ_0, respectively; the vertical and azimuth LOS angle are θ_{0L} and ψ_{0L}, respectively; the velocity of the ith follower missile is V_{mi}; the trajectory inclination angle and the trajectory deflection angle are θ_i and ψ_i, respectively; and the vertical and azimuth LOS angles are θ_{0i} and ψ_{0i}, respectively.

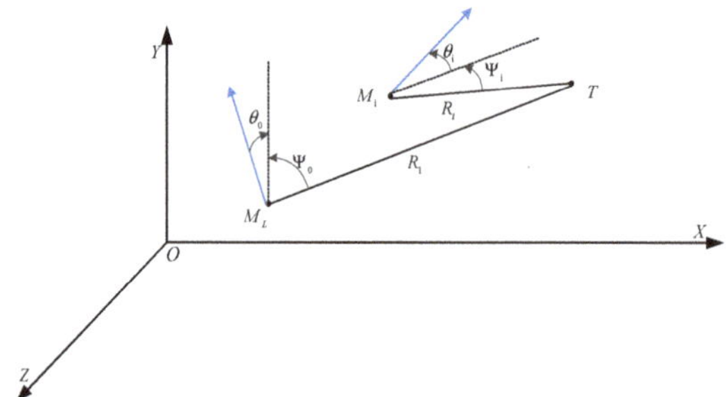

Figure 2. Relative motion relationship of the leader-follower missiles.

In the later stage of the missile movement, the guidance of the missiles makes θ_0 and θ_i tend to zero. At this time, the curvature of the trajectory curve of the leader missile and the follower missiles tends to be consistent, that is, the relationships $r_0/V_{m0} = r_i/V_{mi}$ and $\psi_0 = \psi_i$ (or $r_0/V_{m0} = r_i/V_{mi}$ and $\psi_0 = -\psi_i$) are satisfied, and the leader missile and follower missiles reach and hit the target simultaneously. The important aspect is that this strategy can make a same curvature of the leader missile and follower missiles relative to the velocity. Under this condition, the flight of the missile consumes the same time. Hence, the leader-follower strategy can also be suitable for different missile speeds. The leader missile adopts an augmented proportional guidance method as:

$$A_{z0} = K_{01}|\dot{r}_0||\dot{\theta}_{L0}|/\cos\theta_0 \tag{3}$$

$$A_{y0} = -\frac{K_{01}|\dot{r}_0||\dot{\theta}_{L0}\sin\theta_0\sin\psi_0}{\cos\theta_0\cos\psi_0} - \frac{K_{02}|\dot{r}_0||\dot{\psi}_{L0}\cos\theta_{L0}}{\cos\psi_0} \tag{4}$$

In order to quickly make the attitude angle θ_0 of the leader missile tend towards zero, a larger value of K_{01} needs to be chosen. The follower missiles also adopt an augmented proportional guidance method on the pitch channel:

$$A_{zi} = K_{i1}|\dot{r}_i||\dot{\theta}_{Li}|/\cos\theta_i \tag{5}$$

In order to quickly make the attitude angle θ_i of the leader missile tend towards zero, a larger value of K_{i1} needs to be chosen. Next, according to the position relationship between the leader missile and the follower missiles, the lateral acceleration A_{yi} of the leader missile is designed so that $r_i/V_{mi} \to r_0/V_{m0}$ and $\psi_i \to \pm\psi_0$ are satisfied during the guidance process.

3.2. Cooperative Strategy of Follower Missiles

The remaining time error between the leader missile and follower missiles is defined as follows:

$$\zeta = \frac{r_0}{V_{m0}} - \frac{r_i}{V_{mi}} \tag{6}$$

By deriving the remaining time error with respect to time, we can get:

$$\dot{\zeta} = \cos\theta_i \cos\psi_i - \cos\theta_0 \cos\psi_0 \tag{7}$$

According to Equation (7), A_{yi} indirectly controls ζ by controlling ψ_i. Therefore, the control system can be divided into two sub-control systems according to direct control ψ_i and indirect control ζ, namely, a nonlinear slow sub-system and a nonlinear fast sub-system. Therefore, the dynamic design method of time-scale separation is used to design the control instructions A_{yi}. The desired dynamics of the slow sub-system is expressed as:

$$\dot{\zeta}_{des} = -K_R\zeta \tag{8}$$

where K_R is the bandwidth of the slow sub-system.

If $\dot{\zeta} = \dot{\zeta}_{des}$ is obtained through the control, considering the command value of ψ_i is $\psi_i{}^c$, then:

$$\cos\psi_i{}^c = \frac{\cos\theta_0 \cos\psi_0 - K_R\zeta}{\cos\theta_i} \tag{9}$$

Because the value range of ψ_i is $(-\pi/2, \pi/2)$, $\psi_i{}^c$ is within the value range. Combining the above equation and the value range, the value range of KR can be obtained:

$$\frac{\cos\theta_0 \cos\psi_0 - \cos\theta_i}{\zeta} \le K_R \le \frac{\cos\theta_0 \cos\psi_0}{\zeta} \tag{10}$$

if

$$K_R = \frac{\cos\theta_0 \cos\psi_0 - \cos\theta_i + c_1\frac{\zeta}{\zeta+c_2}\cos\theta_i}{\zeta} \tag{11}$$

where c_1, c_2 are constants and satisfy the relationship $0 < c_1, c_2 < 1$.

Herein, $\psi_i{}^c$ can be expressed as:

$$\psi_i^c = f_\psi \times \arccos\left(\frac{\cos\theta_0 \cos\psi_0 - K_R\zeta}{\cos\theta_i}\right) \tag{12}$$

where $\psi_i \in [-\pi/2, 0)$, $f_\psi = -1$; $\psi_i \in [0, \pi/2]$, $f_\psi = 1$.

Taking the derivative of Equation (12) with respect to time, we have:

$$\dot{\psi}_i^c = f_{\dot{\psi}} \times \left(\frac{K_R \cos\psi_i - \dfrac{\cos\psi_i^c \sin\theta_i}{\cos\theta_i} - }{\cos\theta_i\sqrt{1-(\cos\psi_i^c)^2}} K_R \cos\theta_0 \cos\psi_0 - \dot{\theta}_0 \sin\theta_0 \cos\psi_0 - \dot{\psi}_0 \cos\theta_0 \sin\psi_0 \right) \tag{13}$$

If $\psi_i \in [-\pi/2, 0)$, $f_{\dot{\psi}} = -1$; $\psi_i \in [0, \pi/2]$, $f_{\dot{\psi}} = 1$.

In order to make ξ satisfy the relation in Equation (8), ψ_i should quickly converge to near its command value ψ_i^c. Therefore, the following fast dynamics sub-system is designed, and its specific expression is given as:

$$\dot{\psi}_{i,des} - \dot{\psi}_i^c = -K_\psi(\psi_i - \psi_i^c)$$ (14)

where K_ψ is the bandwidth of the fast sub-system. Under the control $\dot{\psi}_i = \dot{\psi}_{i,des}$ united with Equation (4), the final designed maneuvering control command in the turning plane can be obtained as:

$$A_{yi} = \frac{V_{mi}^2}{r_i}\sin\psi_i(1 - \cos\theta_i\sin\theta_i\cos\psi_i\tan\theta_{Li}) - V_{mi}\cos\theta_i\dot{\psi}_i^c + K_\psi V_{mi}\cos\theta_i(\psi_i - \psi_i^c)$$ (15)

Equation (14) can be satisfied by maneuvering control commands on the lateral plane, so that $\psi_i \to \psi_i^c$ is obtained; ψ_i^c can make the slow dynamics sub-system meet Equation (8). Under this condition, the time cooperative control of multiple missiles can be realized. Longitudinal control commands A_{z0} and A_{zi} can make $\theta_0 \to 0$ and $\theta_i \to 0$, along with the control of the yaw channel, $\xi \to 0$, so as to realize $\psi_i \to \psi_0$ (or $\psi_i \to -\psi_0$). The control flow chart for cooperative guidance of multiple missiles can be seen form Figure 3.

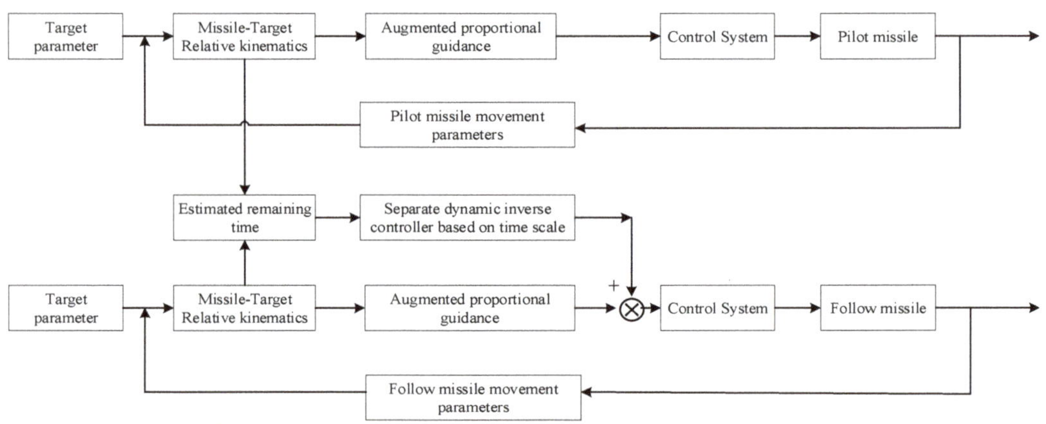

Figure 3. Control flow chart for cooperative guidance of multiple missiles.

4. Cooperative Prediction Guidance Based on Distributed Model Prediction Control (DMPC) for Multiple Missiles

4.1. Introduction to DMPC

The application of the MPC method in a multi-agent cluster system has three structures: centralized, decentralized, and distributed MPC. As shown in Figure 4, the centralized MPC considers the overall performance of the cluster as an optimization index. In each sampling period, a central model predictor performs the rolling optimization on the overall control performance index of the cluster system to obtain the control sequence set with the best overall performance of the cluster system. Each sub-system of the decentralized and distributed MPC has an independent model predictor, which reduces the dimensionality of the system. It also reduces the amount of calculation for the online optimization of each sub-controller and increases the reliability of the system. The difference between the distributed and decentralized MPC is that there is no information interaction between the sub-controllers of the decentralized MPC. As a result, it cannot control the problem of the multi-agent system with coupling between the sub-systems.

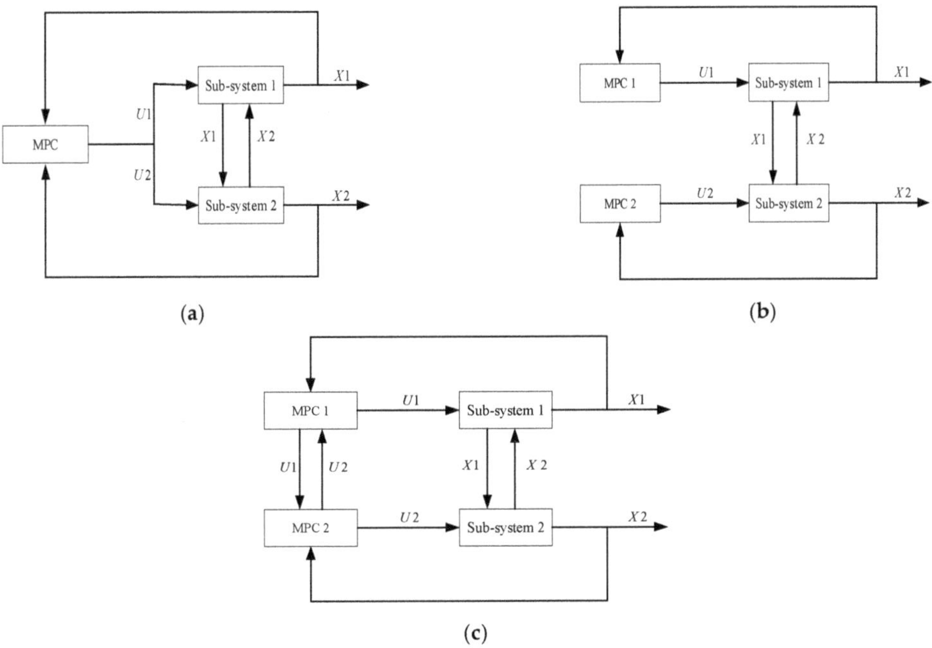

Figure 4. Classification of MPC: (**a**) Centralized MPC; (**b**) Decentralized MPC; (**c**) Distributed MPC.

From the above-mentioned missile motion model, the motion equation of each missile can be expressed as:

$$\dot{z}_l(t) = f_i(z_i(t), u_i(t)), t \geq t_0, z_i(0) = z_0 \tag{16}$$

Therefore, the motion equations of its adjacent missiles can be defined as follows:

$$\dot{z}_{-l}(t) = f_{-i}(z_{-i}(t), u_{-i}(t)), t \geq t_0 \tag{17}$$

The prediction time domain is defined as $T_p \in (0, \infty), s \in [t_c, t_c + T_p]$. The control moment of the receding horizon control is defined as $t_c = t_0 + \delta_c, c \in \{0, 1, 2, \cdots\}$, where δ is the control interval and satisfies $\delta \in (0, T_p]$, $\hat{z}_i(s; t_c)$, $\hat{u}_i(s; t_c)$ is the estimated state sequence and control sequence, respectively, while $z_i^p(s; t_c)$ and $u_i^p(s; t_c)$ are the predicted state sequence and control sequence, respectively. For any time, t_c, the optimal control sequence of the *i*th missile can be expressed as:

$$u_i^*(s; t_c) = \operatorname{argmin}_{u_i^*(s; t_c)} \mathcal{J}_i(z_i(s; t_c), z_{-i}(s; t_c), u_i(s; t_c)) \tag{18}$$

The cost function can be expressed as:

$$\mathcal{J}_i = \int_{t_c}^{t_c + T_p} \mathrm{F}_i\left(z_i^p(s; t_c), \hat{z}_{-i}(s; t_c), u_i^p(s; t_c)\right) ds + \Phi_i\left(z_i^p(t_c + T_p; t_c)\right) \tag{19}$$

where $\mathcal{J}_i$ is the operation function and Φ is the final state function. The following constraints are satisfied:

$$\dot{z}_i^p(s; t_c) = f_i\left(z_i^p(s; t_c), u_i^p(s; t_c)\right) \tag{20}$$

$$\dot{\hat{z}}_{-i}(s; t_c) = f_{-i}(\hat{z}_{-i}(s; t_c), \hat{u}_{-i}(s; t_c)) \tag{21}$$

$$z_i^p(s; t_c), \hat{z}_{-i}(s; t_c) \in Z \tag{22}$$

$$u_i^p(s; t_c), \hat{u}_{-i}(s; t_c) \in U \tag{23}$$

According to the definition of the motion equations of the adjacent missiles, $s \in [t_c, t_c + T_p]$ in the prediction time domain and the estimated control sequence $\hat{u}_{-i}(s; t_c)$ can be obtained by:

$$\hat{u}_{-i}(s; t_c) = \{\hat{u}_j(s; t_c)\}, j \in N_i \tag{24}$$

In the prediction time domain $s \in [t_c, t_c + T_p]$, the estimated control sequence $\hat{u}_j(s; t_c)$ consists of two parts. In the time domain $s \in [t_c, t_{c-1} + T_p]$, the first part is the optimal control sequence $u_j^*(s; t_c)$ of the previous prediction time domain $s \in [t_{c-1}, t_{c-1} + T_p]$, while in the time domain $s \in [t_{c-1} + T_p, t_c + T_p]$, the second part is the optimal control sequence derived from $u_j^*(s; t_{c-1})$ at the time $s = t_{c-1} + T_p$. The expression of $\hat{u}_j(s; t_c)$ is as follows:

$$\hat{u}_j(s; t_c) = \begin{cases} u_j^*(s; t_{c-1}) & s \in [t_c, t_{c-1} + T_p) \\ u_j^*(t_{c-1} + T_p; t_{c-1}) & s \in [t_{c-1} + T_p, t_c + T_p] \end{cases} \tag{25}$$

4.2. Algorithm for Cooperative Guidance Considering Impact Time and Angle Constraints

The framework idea of the distributed MPC is as follows: At each control moment t_c, the current missile control variable is initialized with the optimal control amount of the previous prediction period $[t_{c-1}, t_{c-1} + T_p]$. After that, it exchanges the information with the adjacent missiles and receives the estimated value of the control variable at the moment t_{c-1}. Subsequently, it sends out the estimated value of its own control variable at the moment t_{c-1} to estimate the state of the adjacent missile in the prediction period $[t_c, t_c + T_p]$ and evaluates the cost function $\mathcal{J}_i$ in the prediction period $[t_c, t_c + T_p]$ and uses the particle swarm optimization (PSO) algorithm to solve the optimal control variable sequence in the prediction period $[t_c, t_c + T_p]$. Based on the rolling optimization idea, the optimal control amount of the first update interval is taken to obtain the optimal control sequence, and the missile state at $[t_c, t_{c+1}]$ is updated to enter the next control moment t_{c+1}. The solution process of the estimated value is shown in Figure 5.

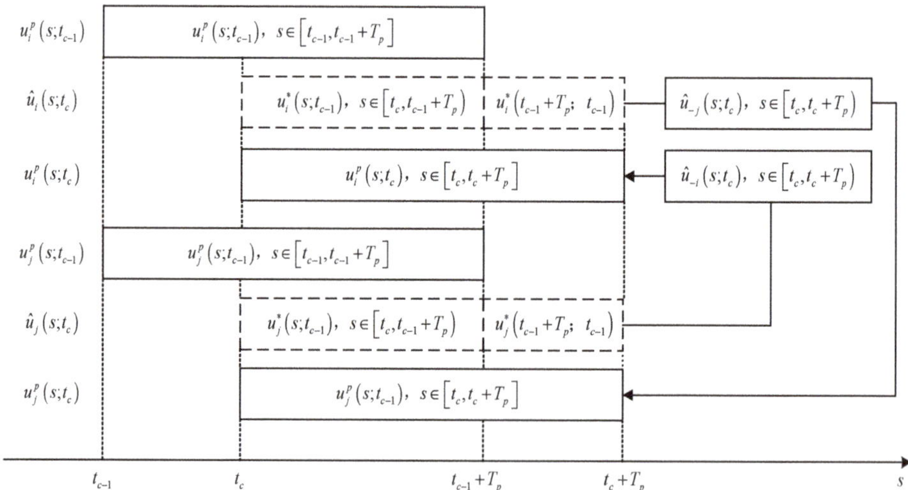

Figure 5. Solution process of the predicted state value.

The basic principle of the PSO-based distributed cooperative model prediction framework for multiple missiles is shown in Figure 6. The cost function of the cooperative control of missiles includes two parts, namely, the operation function, F, and the final state function, Φ:

$$J_i = \int_{t_c}^{t_c+T_p} F_i\left(z_i^p(s;t_c), \hat{z}_{-i}(s;t_c), u_i^p(s;t_c)\right)ds + \Phi_i\left(z_i^p(t_c + T_p;t_c)\right) \tag{26}$$

The operation function is defined as:

$$F_i\left(z_i^p(s;t_c), \hat{z}_{-i}(s;t_c), u_i^p(s;t_c)\right) = \alpha \sum_{j \in N_i} \|t_{go,i}^p(s;t_c) - \hat{t}_{go,j}(s;t_c)\|^2 + (1-\alpha)\|u_i^p(s;t_c)\|^2 \tag{27}$$

The final state function is defined as:

$$\Phi_i\left(z_i^p(t_c + T_p;t_c)\right) = \beta\|r_i(t_c + T_p;t_c)\|^2 + \gamma\|\theta(t_c + T_p;t_c) - \theta_{expect} + \psi(t_c + T_p;t_c) - \psi_{expect}\| \tag{28}$$

where α, β, γ are the weight constants, $\|\cdots\|$ represents the modulus of the space vector, N_i is the set of missiles other than missile i, r_i is the distance between the missile and the target, $t_{go,i}^p$ is the predicted time of arrival, and $\hat{t}_{go,i}$ is the estimated time of arrival, which can be solved by the following equation:

$$\hat{t}_{go,j}(s;t_c) = \frac{\hat{r}(s;t_c)}{v_j}, j \in N_i \tag{29}$$

where $\hat{r}(s;t_c)$ is the estimated distance between the missile and the target at the moment t_c, θ_{expect} and ψ_{expect} represent the desired pitch angle and trajectory deflection angle that are used to control the angle of the missile at the terminal landing point to achieve the impact angle constraints.

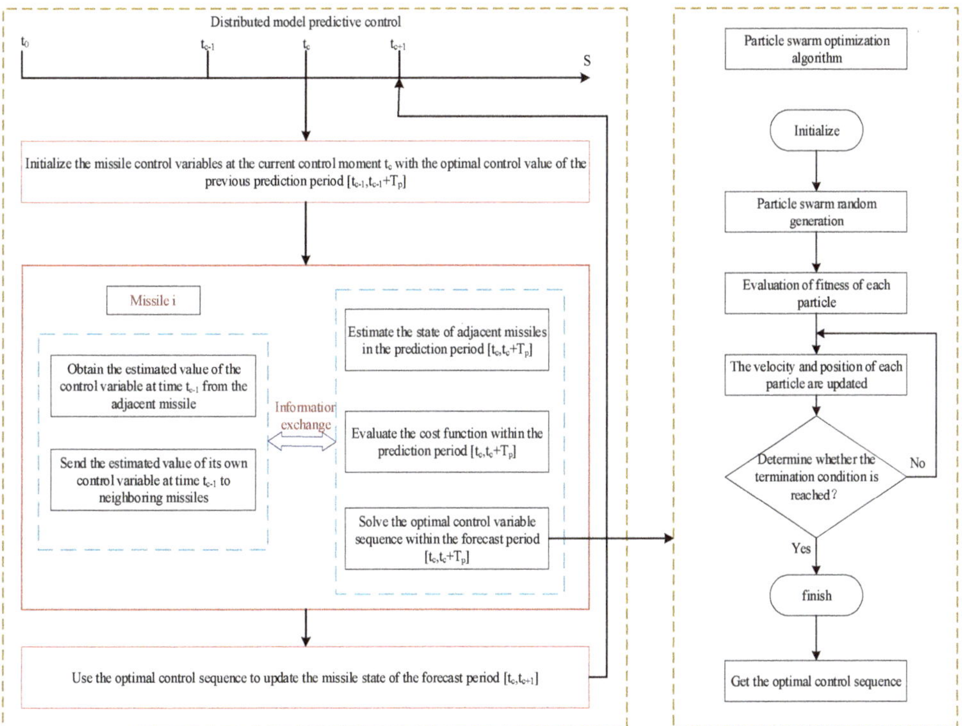

Figure 6. PSO-DMPC model prediction framework.

5. Verification by Simulations

Let M1, M2, M3, and M4 denote four missiles participating in cooperative operations. The target is in the horizontal plane, and its initial position is at the origin of the coordinates. Table 1 lists the initial parameters of the proposed virtual leader missile (denoted as M) and follower missiles.

Table 1. Initial parameters of the virtual leader and the follower missiles.

Name		Initial Position			Velocity	Ballistic Angle	Ballistic Drift Angle
Variable		x	y	z	Vm	θ	ψ
Unit		m	m	m	m/s	(°)	(°)
Virtual Pilot Missile		4000	1800	3000	133	45	45
	M1	−3000	1500	−3800	155	45	45
Follow	M2	−3500	1200	−2500	140	40	40
Missile	M3	3500	800	2000	125	10	30
	M4	−3500	1600	2500	140	−15	20

Let us consider a case where a stationary target, a straight moving target, and a snake-like maneuvering target is attacked (specific movement of the target is shown in Table 2).

Table 2. Initial parameters of the virtual leader missile and follower missiles.

Type	Target Motion State		
Target 1		Static target	
Target 2	$v_t = 10 \text{ m/s}$	$a_t = 0 \text{ m/s}^2$	$\psi_{vt} = 45°$
Target 3	$v_t = 10 \text{ m/s}$	$y_t = 60 \sin(x_t/100)$	$K_{01} = K_{i1} = 10$

The control coefficients of the augmented proportional guidance method adopted by the five missiles are $K_{01} = K_{i1} = 10$, $K_{02} = 5$; the parameters are $c_1 = 0.7$, $c_2 = 0.9$; the bandwidth of the fast dynamics sub-system is $K_\psi = 5$; and the overload limit of the missile is 8 g. The impact time during the attack on a fixed target is shown in Table 3. According to the above design strategy, under the premise of adopting the augmented proportional guidance method, the impact time of the leader missile is greater than that of each follower missile. Therefore, the use of this scheme can provide a sufficient flight time adjustment margin for the follower missiles.

Table 3. Impact time of APN guidance.

Target Type	Virtual Pilot Missile		Follow Missile			
	Unit	M	M1	M2	M3	M4
Target 1	s	41.66	33.92	32.85	33.25	32.93
Target 2	s	39.48	41.36	36.58	33.39	36.54
Target 3	s	39.82	39.95	34.67	34.01	35.31

In the adjustable range, in order to enhance the ability of attack and damage on moving targets in a horizontal plane, the ideal impact angles $(\theta_m^*, \psi_{vm}^*)$ of M1, M2, M3, and M4 are designated as $(50°, -20°)$, $(44°, -34°)$, $(-30°, 0°)$, and $(35°, -20°)$, respectively. The ideal impact time t^* is determined based on the flight time of the leader missile attacking different targets. The simulation results of cooperative attack by missiles on targets 1, 2, and 3 are shown in Figures 7–11.

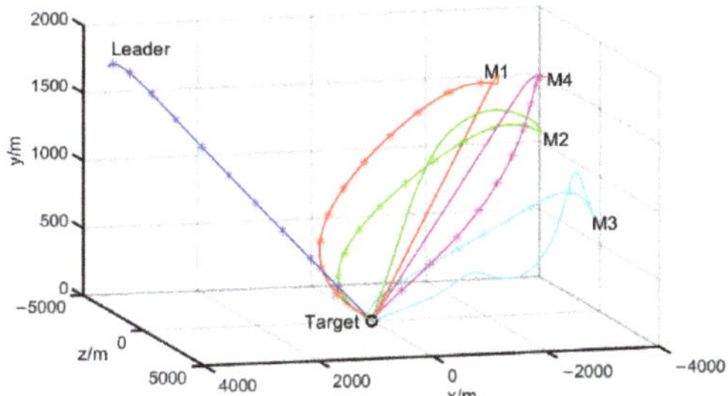

Figure 7. Flight trajectory curve of the cooperative attack on stationary target 1.

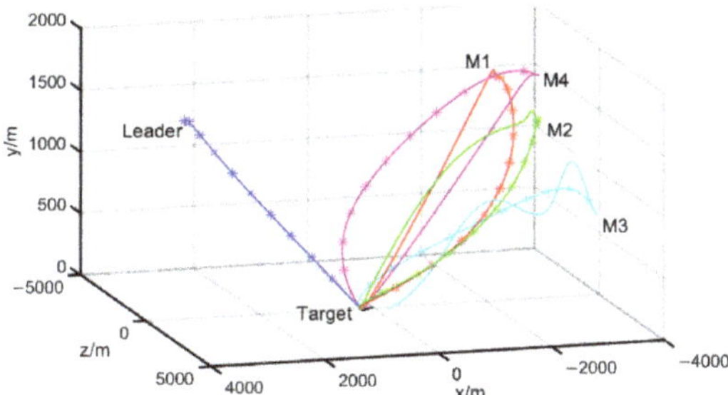

Figure 8. Flight trajectory curve of the cooperative attack on target 2.

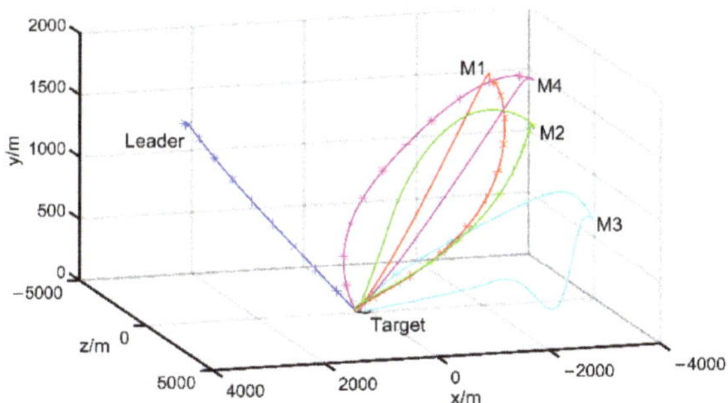

Figure 9. Flight trajectory curve of the cooperative attack on target 3.

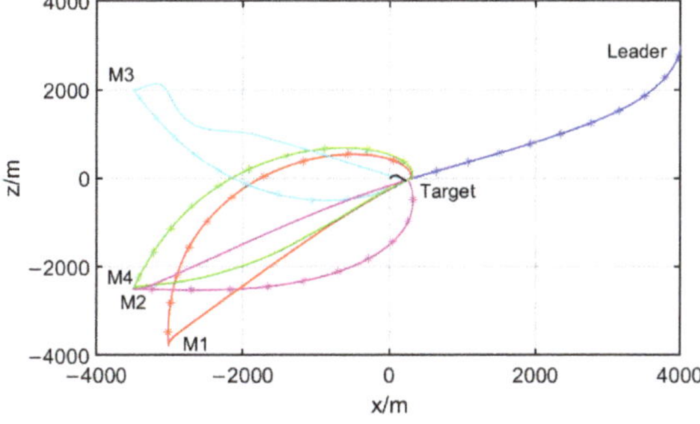

Figure 10. Projection of flight trajectory curve of cooperative attack on target 3 in the XOZ plane.

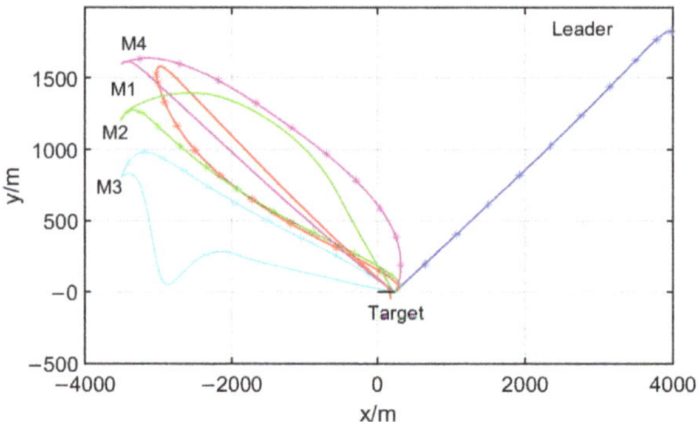

Figure 11. Projection of flight trajectory curve of cooperative attack on target 3 in the XOY plane.

As shown in Figures 7–9, the cooperative strategy based on the tracking of the missile-target distance could well constrain the missile-target distance during the flight of multiple projectiles. As a result, the follower missiles converged to the same value as the leader missile. The MPC-based cooperative guidance algorithm optimizes the trajectory to meet the terminal impact angle constraints, based on the impact time of the cooperative missile group. From Figures 10 and 11, it can be observed that the MPC-based cooperative guidance law can make multiple missiles participating in cooperative operations hit the target with separately specified impact angles at a specified time. The miss distance meets the requirements and is less than that of the cooperative time guidance. Additionally, the strike effect is more accurate. Due to the limited adjustable time and angle of the missiles under different position conditions, the omni-directional saturation attack of the multiple missiles can be realized through reasonable state estimation, mission planning, and assignment of multiple terminal guidance missiles. The simulation results verified the effectiveness of this method in attacking the stationary and maneuvering targets. From Figures 7–10, it can be observed that the trajectory curve optimized by the MPC algorithm is smoother and easier to implement.

By tracking the distance between the missile and the target, each follower missile achieved a cooperative impact time with the virtual leader missile. When hitting target 1, target 2, and target 3, the cooperative impact time of the four missiles was the same as the impact time of the virtual lead missile, which was 41.66 s, 41.28 s, and 41.82 s, respectively. However, the impact angle did not meet the requirements at this time, and the iterative optimization of the guidance instructions through the MPC algorithm achieved the impact angle constraints at the terminal. In the simulation example, the impact angle constrained by each follower missile and the terminal impact angle are given by Figures 12–14, when different guidance methods were used to attack an unfixed target.

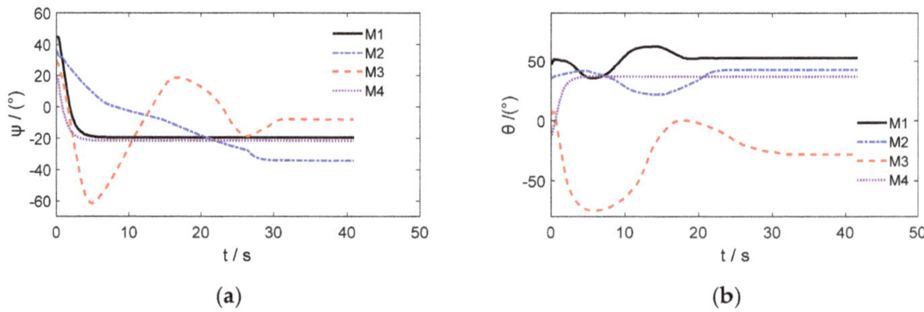

Figure 12. Velocity-inclination change curve of each follower missile during the cooperative attack on target 1 based on the MPC method: (**a**) Trajectory deflection curve; (**b**) Trajectory inclination curve.

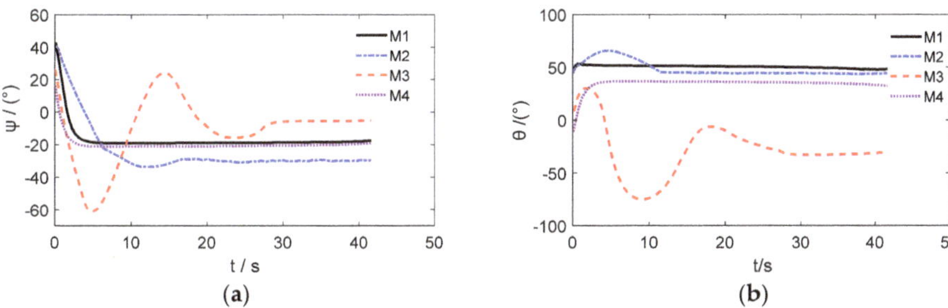

Figure 13. Velocity-inclination change curve of each follower missiles during cooperative attack on target 2 based on the MPC method: (**a**) Trajectory deflection curve; (**b**) Trajectory inclination curve.

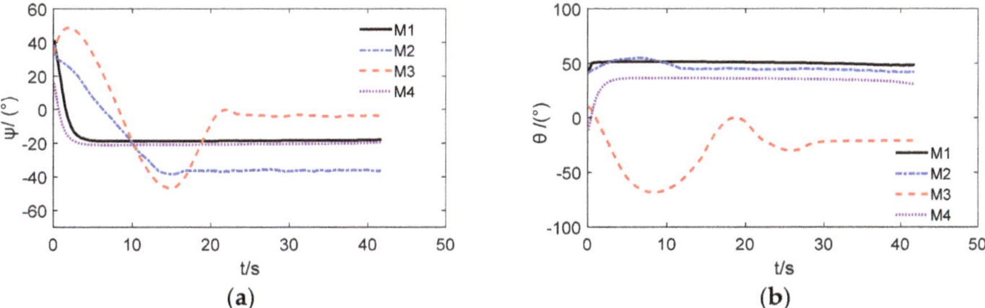

Figure 14. Velocity-inclination change curve of each follower missiles during the cooperative attack on target 3 based on the MPC method: (**a**) Trajectory deflection curve; (**b**) Trajectory inclination curve.

Figures 12–14 show the changing curves of the impact angle when attacking a stationary target, a uniformly moving target, and a maneuvering target by the MPC method. Using the MPC method can not only realize the cooperative time but can also satisfy the impact angle constraints. Both the time cooperative guidance and the MPC-based guidance can make four missiles hit the target at the same time, but the former could only constrain the impact time of four missiles, while the latter achieved a multi-directional cooperative attack with designated angles on the target. This verified the superior guidance performance of the MPC-based 3D cooperative guidance law with multiple constraints. Figure 15 shows the tracking curve of the missile-target distance in the time cooperative strategy with the designed parameters. Figure 16 shows the curve of the remaining time error with the time obtained by the MPC method. Tracking the missile-target distance of the leader missile achieves a cooperative impact time between the follower missiles and the leader missile, but it does not meet the impact angle constraints. Therefore, the MPC method was used to iteratively optimize the control commands to meet the impact angle constraints.

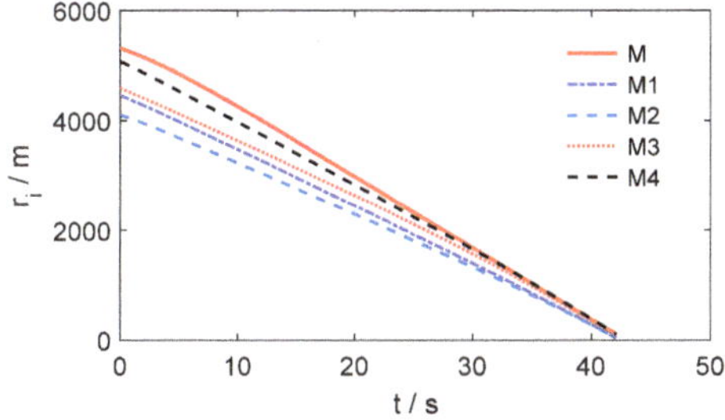

Figure 15. Tracking of the missile-target distance.

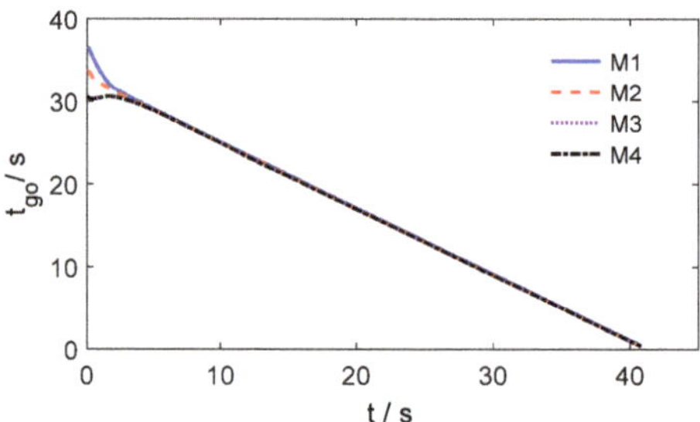

Figure 16. Remaining time error of MPC.

The performance of the MPC-based 3D cooperative guidance law and time cooperative guidance law are shown in Figures 17 and 18, respectively. It can be observed that, when the missile flies with the time cooperative guidance law, although the impact time was similar to the flight time of the leader missile, there was still a significant gap between the terminal attack attitude angle and the ideal impact angle. After iterative optimization using the MPC algorithm, both the trajectory inclination θ_{0L} and the trajectory deflection angle ψ_{0L} achieved the impact angle constraints within the specified time. The MPC algorithm is based on the principle of optimization and was designed according to performance indicators to minimize the total energy in the guidance process under the condition of meeting terminal constraints. In every iteration of the MPC algorithm, the deviation at the terminal of the trajectory was evenly distributed to each step of the entire trajectory by a controlled adjustment. Therefore, the control command exhibited a smoother transition than the initial control command, and the amplitude was smaller, which was easier to implement in engineering.

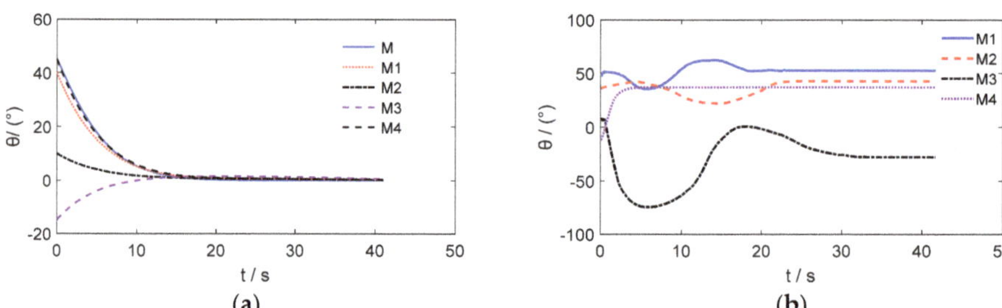

Figure 17. Comparison of the trajectory inclination curves of each follower missile between the proposed guidance scheme and the time cooperative guidance scheme: (**a**) Time cooperative guidance law; (**b**) MPC-based cooperative guidance law.

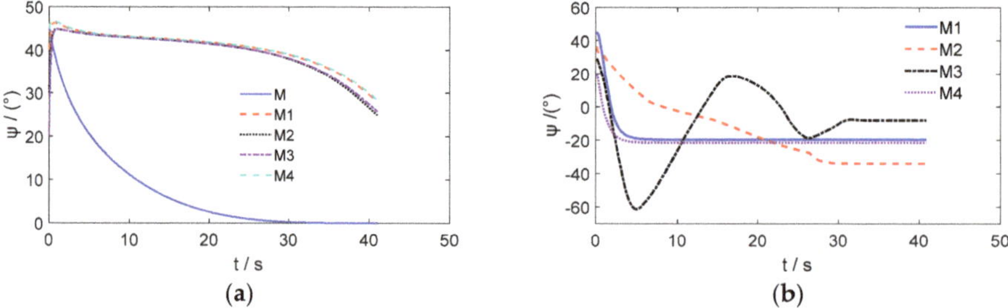

Figure 18. Comparison of the trajectory deflection curves of each follower missile between the proposed guidance scheme and the time cooperative guidance scheme: (**a**) Time cooperative guidance law; (**b**) MPC-based cooperative guidance law.

From Figures 19 and 20, it can be observed that the time cooperative guidance strategy required greater control capabilities, especially at the terminal. In order to achieve the consistency requirements of the time cooperative guidance, a greater overload is required, and the required overload exceeded that which the missile can provide. Therefore, this method has very high requirements for the maneuverability of the missiles. In contrast with the MPC-based cooperative strategy, the control amount can be controlled within a certain range through the strategy of rolling optimization, and the requirements for the maneuverability of the missile are not high. Furthermore, it can not only achieve the time cooperation, but also achieve the requirements of impact angle at the terminal. Thus, it proved to be a better cooperative guidance strategy.

The results demonstrated that the cooperative strategy of tracking the distance between the missile and the target based on the dynamic inverse design was very effective. The MPC algorithm added the impact angle constraints based on the impact time constraints, thereby achieving good effects and a better guidance performance, along with an easy implementation of the instructions.

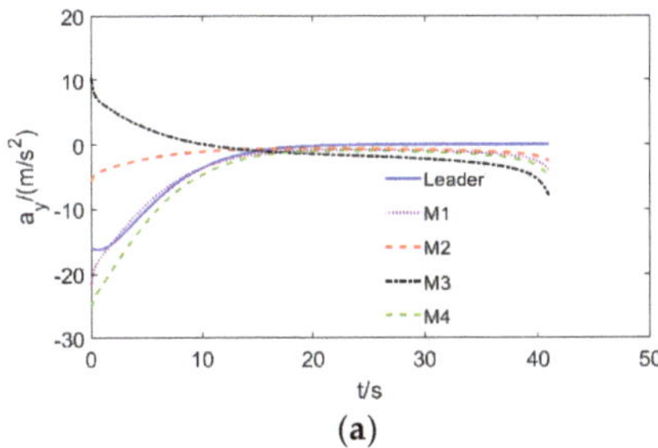

(a)

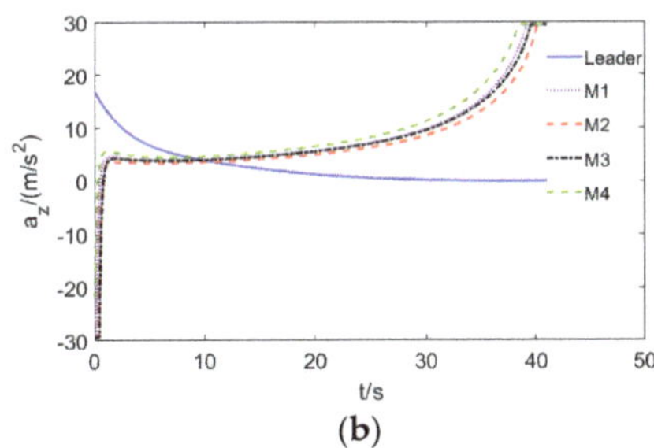

(b)

Figure 19. Normal acceleration curves obtained by the time cooperative guidance algorithm: (**a**) Normal acceleration curve of the dive plane; (**b**) Normal acceleration curve of turning plane.

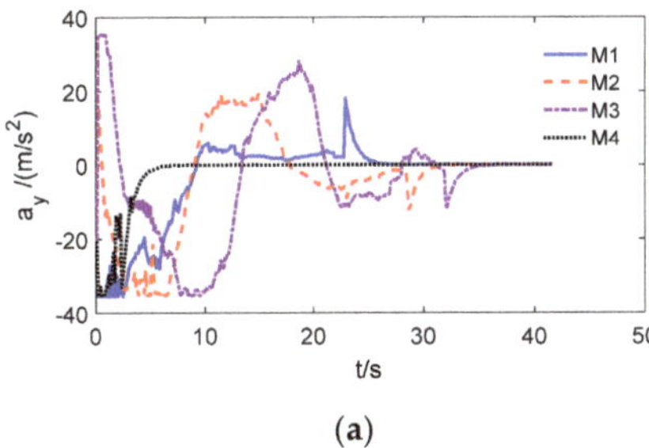

(a)

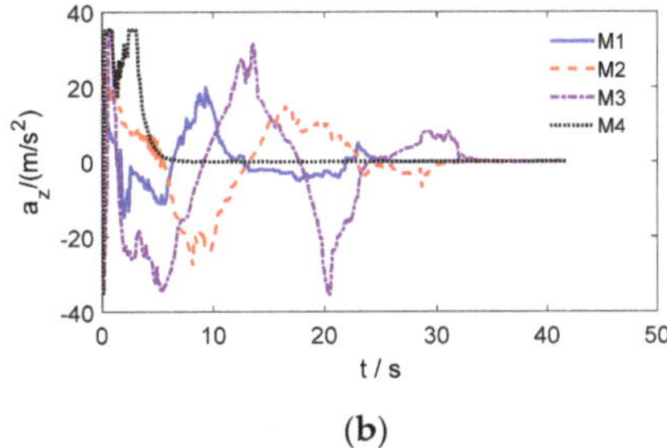

(b)

Figure 20. Normal acceleration curves obtained by the MPC guidance algorithm: (**a**) Normal acceleration curve of the dive plane; (**b**) Normal acceleration curve of turning plane.

6. Conclusions

In the present study, we proposed a three-dimensional cooperative terminal guidance strategy for cruise missiles under multiple constraints, introduced the concept of virtual leader missiles, and designed a cooperative strategy for the impact time of virtual leader and follower missiles. On this basis, using the model predictive control method, a three-dimensional cooperative guidance law was presented that could simultaneously control the multiple missiles with impact time and impact angle constraints on the premise of meeting the requirements of the miss distance. The algorithm implementation was carried out, and the effectiveness of the algorithm was verified using simulation studies. The robustness of the algorithm and the solution time of the algorithm both need further study in order to improve the operating efficiency of the algorithm and realize real-time calculations.

Author Contributions: Conceptualization, M.C. and X.C.; methodology, M.C. and Z.Z.; validation, Z.Z. and Z.L.; formal analysis, X.C. and Z.Z. data curation, Z.L.; writing—original draft preparation, Z.Z. and Z.L.; writing—review and editing, M.C. All authors have read and agreed to the published version of the manuscript.

Funding: This research received no external funding.

Institutional Review Board Statement: Not applicable.

Informed Consent Statement: Not applicable.

Data Availability Statement: Not applicable.

Conflicts of Interest: The authors declare no conflict of interest.

References

1. Zhao, Q.; Dong, X.; Liang, Z.; Bai, C.; Chen, J.; Ren, Z. Distributed cooperative guidance for multiple missiles with fixed and switching communication topologies. *Chin. Aeronaut.* **2017**, *30*, 1570–1581. [CrossRef]
2. Nikusokhan, M.; Nobahari, H. Closed-form optimal cooperative guidance law against random step maneuver. *IEEE Trans. Aerosp. Electron. Syst.* **2016**, *52*, 319–336. [CrossRef]
3. Chen, P.; Xing, L.; Wu, S.-T.; Li, M. Concensus problems in distributed cooperative terminal guidance time of multi-missiles. *Control Decis.* **2010**, *25*, 1557–1561. (In Chinese)
4. Wang, L.; Ma, G.; Jiao, Y. Guidance law design of multi-aircraft formation cooperative interception with Angle constraint. *Nav. Aeronaut. Astronaut. Univ.* **2018**, *33*, 289–296. (In Chinese)
5. Li, G.; Yu, Z.; Zhang, Y. Cooperative guidance law with angle constraint to intercept maneuvering target. *Syst. Eng. Electron.* **2019**, *41*, 626–635. (In Chinese)
6. Zhao, B.; Huang, X.; Zhou, J.; Guo, Y. Multi-missile Distributed LOS Cooperative Guidance Law Designed Based on Sliding Mode Control. *Air Space Def.* **2020**, *3*, 16–23. (In Chinese)
7. Wang, Q.; Liu, M.; Ren, J.; Wang, T. Overview of Common Algorithms for UAV Path Planning. *J. Jilin Univ. (Inf. Sci. Ed.)* **2019**, *37*, 58–67. (In Chinese)
8. Wang, X.; Zheng, Y.; Lin, H. Integrated guidance and control law for cooperative attack of multiple missiles. *Aerosp. Sci. Technol.* **2015**, *42*, 1–11. [CrossRef]
9. Ates, U.H. Nonlinear impact angle control guidance law for stationary targets. In Proceedings of the AIAA Guidance, Navigation, and Control Conference, San Diego, CA, USA, 4–8 January 2016; pp. 2016–2112.
10. Gu, Y. Guidance Law of Multi-Missiles Formation Fight. *Mod. Def. Technol.* **2014**, *42*, 51–56. (In Chinese)
11. Wang, X.; Wang, J. Partial integrated guidance and control with impact angle constraints. *J. Guid. Control Dyn.* **2014**, *37*, 644–657. [CrossRef]
12. Tekin, R.; Erer, K.S. Switched-gain guidance for impact angle control under physical constraints. *J. Guid. Control Dyn.* **2015**, *38*, 205–216. [CrossRef]
13. Jeon, I.S.; Lee, J.I.; Tahk, M. Impact-time-control guidance law for anti-ship missiles. *IEEE Trans. Control Syst. Technol.* **2006**, *14*, 260–266. [CrossRef]
14. Yang, Z.; Lin, D.; Wang, H. Impact time control guidance law with filed-of-view limit. *Syst. Eng. Electron.* **2018**, *38*, 2122–2128. (In Chinese)
15. Zhao, Q.L.; Chen, J.; Dong, X.W.; Wang, R.; Li, Q.D.; Zhang, R. Cooperative guidance law for heterogeneous missiles interecepting hypersonic weapon. *Acta Aeronaut. Astronaut. Sin.* **2016**, *37*, 936–948. (In Chinese)
16. Ye, P.; Zhang, J.; Li, Y.; Qi, G.; Sheng, D. Distributed cooperative guidance for multiple missiles with different line of sight Angle constraints. *Acta Armamentarll* **2019**, *40*, 506–515. (In Chinese)
17. Jeon, I.S.; Lee, J.I. Impact-time-control guidance law with constraints on seeker look angle. *IEEE Trans. Aerosp. Electron. Syst.* **2017**, *53*, 2621–2672. [CrossRef]
18. Chen, X.; Wang, J. Nonsingular sliding-mode control for field-of-view constrained impact time guidance. *J. Guid. Control Dyn.* **2018**, *41*, 1214–1222. [CrossRef]
19. Guo, Y.; Wang, H. Overview of cultural algorithms. *Comput. Eng. Appl.* **2009**, *45*, 41–46. (In Chinese)
20. Cho, N.; Kim, Y. Modified pure proportional navigation guidance law for impact time control. *J. Guid. Control Dyn.* **2016**, *39*, 852–872. [CrossRef]
21. Kang, S.; Kim, H. Differential game missile guidance with impact angle and time constraints. *IFAC Proc. Vol.* **2011**, *44*, 3920–3925. [CrossRef]
22. Erer, K.S.; Tekin, R. Impact time and angle control based on constrained optimal solutions. *J. Guid. Control Dyn.* **2016**, *39*, 2448–2454. [CrossRef]
23. Harl, N.; Balakrishnan, S.N. Impact time and angle guidance with sliding mode control. *IEEE Trans. Control Syst. Technol.* **2012**, *20*, 1436–1449. [CrossRef]

24. Kim, T.H.; Lee, C.H.; Jeon, I.S.; Tahk, M.J. Augmented polynomial guidance with impact time and angle constraints. *IEEE Trans. Aerosp. Electron. Syst.* **2013**, *49*, 2806–2817. [CrossRef]
25. Padhi, R.; Kothari, M. Model predictive static programming: A computationally efficient technique for suboptimal control design. *Int. J. Innov. Comput. Inf. Control* **2009**, *5*, 399–411.
26. Wang, J.; Li, F.; Zhao, J.; Wan, C. Review of multi-missile cooperative guidance law. *Flight Mechan.* **2011**, *29*, 13–18. (In Chinese)
27. Wang, X.; Hong, X.; Lin, H. A Method of Controlling Time and Impact Angle of Multiple-missiles Cooperative Combat. *J. Ballist.* **2012**, *24*, 1–5. (In Chinese)
28. Wang, Z.; Wang, X.; Lin, H. A Cooperative Guidance Law with Constraints of Impact Time and Impact Angle. *J. Ballist.* **2017**, *29*, 1–8. (In Chinese)
29. Song, J.; Song, S.; Xu, S. Multi-missile cooperative guidance law with Angle of attack constraint. *J. Chin. Inert. Technol.* **2016**, *24*, 554–560. (In Chinese)
30. Zhang, C.; Song, J.; Hou, B.; Zahng, M. Cooperative Guidance Law with Impact Angle and Impact Time Constraints for Networked Missiles. *Acta Armamentarll* **2016**, *37*, 431–438. (In Chinese)
31. Li, W.; Li, H.; Wang, L.; Gao, J. Design of Three-dimensional Guidance Law for Missiles Cooperative Engagement with Multiple Constraints. *Aerosp. Control* **2018**, *36*, 12–19. (In Chinese)
32. Slnai, E.; Lovera, M. Magnetic spacecraft attitude control: A survey and some new results. *Control Eng. Pract.* **2005**, *13*, 357–371.
33. Elsisi, M.; Ebrahim, M.A. Optimal design of low computational burden model predictive control based SSDA towards autonomous vehicle under vision. *Int. J. Intell. Syst.* **2021**, *26*, 6969–6987. [CrossRef]
34. Elsisi, M. Optimal design of nonlinear model predictive controller based on new modified multitracker optimization algorithm. *Int. Intell. Syst.* **2020**, *35*, 1857–1878. [CrossRef]
35. Shamaghdari, S.; Nikravesh, S.K.Y.; Haeri, M. Integrated guidance and control of elastic flight vehicle based on Robust MPC. *Int. J. Robust Nonlinear Control* **2015**, *25*, 2608–2630. [CrossRef]
36. Yao, Y.; Yang, B.; He, F.; Qiao, Y.; Cheng, D. Attitude control of missile via fliess expansion. *IEEE Trans. Control Syst. Technol.* **2008**, *16*, 959–970. [CrossRef]

MDPI

Article

Fully Distributed Control for a Class of Uncertain Multi-Agent Systems with a Directed Topology and Unknown State-Dependent Control Coefficients

Zongcheng Liu [1,*], Hanqiao Huang [2,*], Sheng Luo [2], Wenxing Fu [2] and Qiuni Li [1,*]

1   Aeronautics Engineering College, Air Force Engineering University, Xi'an 710038, China
2   Unmanned System Research Institute, Northwestern Polytechnical University, Xi'an 710072, China;
     luosheng611@163.com (S.L.); Wenxingfu@nwpu.edu.cn (W.F.)
*   Correspondence: liu434853780@163.com (Z.L.); cnxahhq@126.com (H.H.); lqnjk1@126.com (Q.L.);
     Tel.: +86-029-84787345 (Z.L.)

Abstract: To address the control of uncertain multi-agent systems (MAS) with completely unknown system nonlinearities and unknown control coefficients, a global consensus method is proposed by constructing novel filters and barrier function-based distributed controllers. The main contributions are as follows. Firstly, a novel two-order filter is designed for each agent to produce informational estimates from the leader, such that a connectivity matrix is not used in the controller's design, solving the difficultly caused by the time-varying control coefficients in a MAS with a directed graph. Secondly, combined with the novel filters, barrier functions are used to construct the distributed controller to deal with the completely unknown system nonlinearities, resulting in the global consensus of the MAS. Finally, it is rigorously proved that the consensus of the MAS is achieved while guaranteeing the prescribed tracking-error performance. Two examples are given to verify the effectiveness of the proposed method, in which the simulation results demonstrate the claims.

Keywords: distributed control; MAS; flight control

Citation: Liu, Z.; Huang, H.; Luo, S.; Fu, W.; Li, Q. Fully Distributed Control for a Class of Uncertain Multi-Agent Systems with a Directed Topology and Unknown State-Dependent Control Coefficients. *Appl. Sci.* **2021**, *11*, 11304. https://doi.org/10.3390/app112311304

Academic Editors: Wei Huang and Dong Zhang

Received: 9 October 2021
Accepted: 16 November 2021
Published: 29 November 2021

1. Introduction

The control of uncertain nonlinear systems has been researched for several decades, such that so many remarkable results have been obtained on this topic [1–9]. However, most of them are for SISO or MIMO systems, and their methods or techniques cannot be directly applying to multi-agent systems, as the information of each agent or subsystems is only available for part of others. According to the topology of information transformation graph, the graph can be divided into undirected and directed graphs. Generally, the consensus control of a MAS with the directed graph is more difficult than the undirected case, since the methods for the directed case are always applicable for the undirected case, but not vice versa.

Recently, some significant progress has been made in the control of a MAS [10–12]. For a linear MAS with undirect graphs, fully distributed adaptive consensus controller is present in [10]. Adaptive asymptotically consensus for an uncertain MAS is achieved in [11], and adaptive asymptotically consensus is achieved in [12] for an uncertain MAS, and so on. However, their methods are only applicable for a MAS with an undirected graph and are in vain for a MAS with a directed graph. For a MAS with a directed graph and constant control coefficients, adaptive consensus for a MAS with system nonlinearities satisfying match conditions is researched in [13] to solve the problem of actuator faults; a fully distributed adaptive consensus control is studied for a MAS with unknown control directions in [14] by using a Nussbaum gain technique; actuator faults in a MAS are considered in [15] with integral chain dynamics; and prescribed performance consensus control for uncertain MAS is investigated in [16]. Though much progress has been made [17–20], it should be noted that there are still some nonnegligible problems to be solved. Firstly, the existing methods require the control coefficients to be constants, or even known, for a MAS with a directed graph. The main difficulty is that the Laplace matrix for a directed graph is asymmetric and thus the selections of control parameters must always resort to adaptive methods, which falls into trouble when the control coefficients are time-varying and unknown. Secondly, to the best of our knowledge, there is no global consensus control method for a MAS with a directed graph and the systems functions thereof completely unknown, except for [21], wherein the unknown system nonlinearities required to satisfy the Lipschitz conditions and control coefficients are one. Universal approximators such as neural networks (NN) or fuzzy logic systems (FLS) have been attempted to solve the consensus control problem of a MAS with completely unknown system nonlinearities [22–24], however, it is well known that these methods are semi-global in the sense that their stabilities depend on the initial conditions of systems and the careful selection of controller parameters. Therefore, NN or FLS-based approaches cannot guarantee the global consensus of the MAS, though they are very favorable to solve the problem of MAS with unknown nonlinearities.

As for the global control of systems with completely unknown nonlinearities, a pioneering work is [25], wherein a low-complexity controller is presented that cannot only achieve global convergence of all the system signals, but which can also guarantee the prescribed performance of tracking error and state errors. In view of the low complexity and strong robustness of this method, much research has been carried on this method for solving different nonlinear control problems [26–30]. By introducing a novel barrier function, a fault-tolerant controller is designed for a class of unknown nonlinear systems in [26]. With consideration to the constraints of system states, a barrier function-based adaptive control method is proposed in [27]. Addressing systems with unknown control direction and system dynamics, a Nussbaum function-based low-complexity control scheme is designed in [28]. As regards asymptotic tracking control for systems with unknown nonlinearities, an universal global low-complexity controller is proposed in [31]. Nevertheless, it is worth mentioning that the global control of a MAS with unknown nonlinearities is still an unsolved problem, since these methods are based on the condition that the desired output for systems are known, but this knowledge cannot be obtained for some agents of a MAS. Moreover, considering the control coefficients of each agent are time-varying functions, these traditional methods will fall into trouble when solving for the consensus control of a MAS with unknown dynamics.

Motived by the above discussion, we investigate the fully distributed control of a MAS with a directed graph, time-varying control coefficients and completely unknown system nonlinearities. The main contributions of this paper are summarized as follows.

(1) To address the time-varying control coefficients of a MAS, a two-order filter is firstly designed for each agent to produce estimates of the signals from the leader, so that an asymmetric Laplace matrix for a directed graph will not be used to design the controller for each agent of the MAS, by which the difficulty of control design is solved.

(2) To address the completely unknown system nonlinearities in MAS, barrier functions are used to propose a fully distributed controller by combining novel filters; barrier functions are well-suited to dealing with the effects of unknown system nonlinearities, such that global results are achieved, for the first time, in a MAS with completely unknown system nonlinearities in this paper.

(3) To guarantee the prescribed tracking performance by the proposed controller, such that the consensus of the controlled MAS is rigorously proved and all the closed signals are globally bounded.

2. Problem Statement and Preliminaries

Consider a class of uncertain MAS as follows

$$\begin{cases} \dot{x}_{i,m} = g_{i,m}(\overline{x}_{i,m})x_{i,m+1} + f_{i,m}(\overline{x}_{i,m}) + d_{i,m}(t,\overline{x}_{i,m}), \ m = 1,2\ldots,n-1 \\ \dot{x}_{i,n} = g_{i,n}(\overline{x}_{i,n})u_i + f_{i,n}(\overline{x}_{i,n}) + d_{i,n}(t,\overline{x}_{i,n}) \\ y_i = x_{i,1}, \ for \ i = 1,2,\ldots,N \end{cases} \tag{1}$$

where $\overline{x}_{i,m} = [x_{i,1},x_{i,2},\ldots,x_{i,m}]^T \in R^m$, $y \in R$, $u \in R$, are the states, the control input and the output of the i th subsystem, respectively. The system nonlinearities $f_{i,m}(\cdot), g_{i,m}(\cdot) : R^m \times R^+ \to R$ are unknown continuous functions with respect to $\overline{x}_{i,m}$. $d_{i,m}(t,\overline{x}_{i,m}), m = 1,2\ldots,n$ represent the system uncertainties and external disturbances.

The desired trajectory for the outputs of the subsystems y_d is bounded and only known by part of the N subsystems, with $\dot{y}_d$ being bounded and unknown to all subsystems.

Suppose that the information transmission condition among the group of N subsystems can be represented by a directed graph $G \triangleq (V,E)$, where $V = \{1,\ldots,N\}$ denotes the set of indexes corresponding to each subsystem. The edge $(i,j) \in E$ indicates that subsystem j could obtain information from subsystem i, but not necessarily vice versa. In this case, subsystem j is called a neighbor of subsystem i, and vice versa. Denoting the set of neighbors for subsystem i as $N_i \triangleq \{j \in V : (j,i) \in E\}$. Self-edging (i,i) is not allowed, thus $(i,i) \notin E$ and $i \notin N_i$. The connectivity matrix $A = [a_{ij}] \in R^{N \times N}$ of G is defined as $a_{ij} = 1$ if $(j,i) \in E$ and $a_{ij} = 0$ if $(j,i) \notin E$. An in-degree matrix Δ is introduced, such that $\Delta = diag(\Delta_i) \in R^{N \times N}$ with $\Delta_i = \sum_{j \in N_i} a_{ij}$ being the i th row sum of A. Then, the Laplacian matrix of L is defined as $L = \Delta - A$. Defining $B = diag\{\mu_1, \mu_2, \ldots, \mu_N\}$, where $\mu_i = 1$ means the y_d is accessible directly by subsystem i, and otherwise, we have $\mu_i = 0$. Throughout this paper, the following notations are used. $\| \cdot \|$ is the Euclidean norm of a vector. Letting $a \in R^n$ and $b \in R^n$ be two vectors, then define the vector operator $.*$ as $a.*b = [a(1)b(1),\ldots,a(n)b(n)]^T$. Letting Q be a matrix, $\lambda_{\min}(Q)$ then denotes the minimum eigenvalue of Q.

Assumption 1. *The directed graph G contains a spanning tree, and the desired trajectory y_d is accessible to at least one subsystem, i.e., $\sum_{i=1}^N \mu_i > 0$.*

Assumption 2. *There exist unknown local Lipschitz functions $b_{i,m}(\overline{x}_{i,m})$ such that, for $i = 1,2,\ldots,N$*

$$|d_{i,m}(t,\overline{x}_{i,m})| \leq b_{i,m}(\overline{x}_{i,m}), m = 1,2,\ldots,n \tag{2}$$

Assumption 3. *The unknown control coefficients $g_{i,m}(\overline{x}_{i,m})$ is strictly positive or negative. Without a loss of generality, it is assumed to be strictly positive, namely, for $i = 1,2,\ldots,N$*

$$g_{i,m}(\overline{x}_{i,m}) > 0, m = 1,2,\ldots,n \tag{3}$$

Lemma 1. *(Ref. [17]) Based on Assumption 1, the matrix $(L+B)$ is nonsingular. Defining*

$$\begin{aligned} \theta &= [\theta_1,\ldots,\theta_N]^T = (L+B)^{-1}[1,\ldots,1]^T \\ P &= diag\{P_1,\ldots,P_N\} = diag\left\{\frac{1}{\theta_1},\ldots,\frac{1}{\theta_N}\right\} \\ Q &= P(L+B) + (L+B)^T P \end{aligned} \tag{4}$$

then $\theta_i > 0$ for $i = 1,2,\ldots,N$ and Q is definitely positive.

Remark 1. *In contrast to the methods in [13–16] for a MAS with a directed graph, the control coefficients, $g_{i,m}(\overline{x}_{i,m})$, are time-varying and unknown continuous functions in this paper, which makes the control design much more difficult, since the matrix P in (4) is always unknown and required to be estimated adaptively while the unknown control coefficients $g_{i,1}(\overline{x}_{i,m})$ make P inestimable. To cope with this problem, a novel two-order filter will be given for each agent (shown later).*

Remark 2. *The system nonlinearities, $f_{i,m}(\overline{x}_{i,m})$ and $g_{i,m}(\overline{x}_{i,m})$, are completely unknown functions so that there is little knowledge with which to construct the controller. To deal with this problem, neural networks and fuzzy logic systems have been used to approximate the unknown functions caused by the system nonlinearities $f_{i,m}(\overline{x}_{i,m})$ and $g_{i,m}(\overline{x}_{i,m})$ in [22–24], however, only semi-global results can be obtained by use of these approximators. To construct a distributed controller for a MAS with these unknown system nonlinearities with global consensus is a challenging problem, which is solved by the skillfull cooperation of novel two-order filters and barrier functions in the following.*

3. Design of Distributed Controller and Filters

In this section, a distributed asymptotic tracking controller for a multi-agent system (1) will be designed. To facilitate the control design in distributed manner, design a filter $(q_{i,1}, q_{i,2})$ for each agent i, with $i = 1, \ldots, N$.

3.1. Filters Design

Denote

$$z_{i,j} = \sum_{k=1}^{N} a_{i,k}(q_{i,j} - q_{k,j}) + \mu_i(q_{i,j} - y_d^{(j-1)}), j = 1, 2 \tag{5}$$

Then, design the filters as

$$\begin{cases} \dot{q}_{i,1} = q_{i,2} \\ \dot{q}_{i,2} = v_i \end{cases} \tag{6}$$

with

$$v_i = -c_1 \underline{z}_i - c_0 q_{i,2} - c_0 \mathrm{sgn}(\underline{z}_i) \sum_{j=1}^{2} \hat{F}_{i,j} \tag{7}$$

$$\dot{\hat{F}}_{i,j} = \sum_{k=1}^{N} a_{i,k}(\hat{F}_{k,j} - \hat{F}_{i,j}) + \mu_i(F_j - y_d^{(j-1)}), \; j = 1, 2 \tag{8}$$

where $\underline{z}_i = c_0 z_{i,1} + z_{i,2}$, $y_d^{(0)} = y_d$ and $y_d^{(1)} = \dot{y}_d$, and c_0, c_1 are design parameters chosen as $c_0 \geq 1$ and $c_1 > c_0 + 1$. We then have the following lemma.

Lemma 2. *Consider a closed-loop system consisting of Nfilters (6) satisfying Assumption 1 with local controller (7). The asymptotic consensus tracking of all the filter's outputs to $y_d(t)$ is achieved, i.e., $\lim_{t \to +\infty} |q_{i,1} - y_d(t)| = 0$. Moreover, $q_{i,1}$ and $q_{i,2}$ are bounded.*

Proof (of Lemma 2). Consider the following Lyapunov function

$$V_z = \frac{1}{2} z^T P z + \frac{1}{2\gamma} \sum_{j=1}^{2} \tilde{F}_j^T P \tilde{F}_j \tag{9}$$

where $z = [\underline{z}_1, \underline{z}_2, \ldots, \underline{z}_N]^T$, $\tilde{F}_j = \hat{F}_j - F_j$, $\hat{F}_j = [\hat{F}_{1,j}, \hat{F}_{2,j}, \ldots, \hat{F}_{N,j}]^T$, $F_j = [F_{1,j}, F_{2,j}, \ldots, F_{N,j}]^T$, and $\gamma > 0$ is a constant satisfying $\gamma < \frac{2\lambda_{\min}^2(Q)}{\varphi^2}$ with $\varphi = \|P(L+B)\|$. Denote $z_j = [z_{1,j}, z_{2,j}, \ldots, z_{N,j}]^T$ and $q_j = [q_{1,j}, q_{2,j}, \ldots, q_{N,j}]^T$. Then, we have

$$\dot{z} = (L+B)(c_0 q_2 - c_0 y_d^{(1)} + v - y_d^{(2)}) \tag{10}$$

Using (9) and (10), the time derivative of V_z is

$$\dot{V}_z = z^T P(L+B)\left(-c_1 z - c_0 \sum_{j=1}^{2} \mathrm{sgn}(\underline{z}).*\underline{F}_j\right.$$

$$+c_0 \sum_{j=1}^{2} \varepsilon(\underline{z}).*\underline{F}_j - c_0 y_d^{(1)} - y_d^{(2)}\Big)$$

$$-\frac{1}{\gamma} \sum_{j=1}^{2} \widetilde{F}_j^T P(L+B)\widetilde{F}_j$$

$$\leq -c_1 z^T Q z - \frac{1}{\gamma} \sum_{j=1}^{2} \widetilde{F}_j^T Q \widetilde{F}_j - c_0 \sum_{j=1}^{2} z^T P \Delta \mathrm{sgn}(\underline{z}).*\underline{F}_j$$

$$+c_0 \sum_{j=1}^{2} z^T P A \mathrm{sgn}(\underline{z}).*\underline{F}_j - c_0 \sum_{j=1}^{2} z^T P B \mathrm{sgn}(\underline{z}).*\underline{F}_j$$

$$-\sum_{j=1}^{2} z^T P(L+B)\left(c_0 y_d^{(1)} + y_d^{(2)}\right) + c_0 \sum_{j=1}^{2} \|z\| \|P(L+B)\| \|\widetilde{F}_j\| \tag{11}$$

where $\mathrm{sgn}(\underline{z}) = [\mathrm{sgn}(\underline{z}_1),\dots,\mathrm{sgn}(\underline{z}_N)]^T$.

By noting

$$c_0 \sum_{j=1}^{2} z^T P \Delta \mathrm{sgn}(\underline{z}).*\underline{F}_j = c_0 \sum_{j=1}^{2} F_j \sum_{i=1}^{N} p_i a_{ik} |\underline{z}_i| \tag{12}$$

$$c_0 \sum_{j=1}^{2} z^T P A \mathrm{sgn}(\underline{z}).*\underline{F}_j \leq c_0 \sum_{j=1}^{2} F_j \sum_{i=1}^{N} p_i a_{ik} |\underline{z}_i| \tag{13}$$

$$c_0 \sum_{j=1}^{2} z^T P B \mathrm{sgn}(\underline{z}).*\underline{F}_j = c_0 \sum_{j=1}^{2} F_j \sum_{i=1}^{N} \mu_i p_i |\underline{z}_i| \tag{14}$$

$$\left| \sum_{j=1}^{2} z^T P(L+B)\left(c_0 y_d^{(1)} + y_d^{(2)}\right) \right| \leq c_0 \sum_{j=1}^{2} F_j \sum_{i=1}^{N} \mu_i p_i |\underline{z}_i| \tag{15}$$

$$\sum_{j=1}^{2} \|z\| \|P(L+B)\| \|\widetilde{F}_j\| \leq \lambda_{\min}(Q) \|z\|^2 + \sum_{j=1}^{2} \frac{\varphi^2}{2\lambda_{\min}(Q)} \|\widetilde{F}_j\| \tag{16}$$

we have

$$\dot{V}_z \leq -c_2 \|z\|^2 - \gamma^* \|\widetilde{F}_j\| \tag{17}$$

where $c_2 = \lambda_{\min}(Q)(c_1 - c_0)$, $\gamma^* = \lambda_{\min}(Q)\left(\frac{1}{\gamma} - \frac{\varphi^2}{2\lambda_{\min}^2(Q)}\right)$. It is easily verified that $c_2 > 0$ and $\gamma^* > 0$, therefore, it follows from (17) that $\lim_{t\to+\infty} \|z\| = 0$ and hence $\lim_{t\to+\infty} |q_{i,1} - y_d(t)| = 0$. From the boundedness of V_z and $\|z\|$, the boundedness of $q_{i,1}$ and $q_{i,2}$ are easily obtained. This completes the proof.

Remark 3. *As is seen, a two-order filter is designed to produce a signal $q_{i,1}$ for each agent. Actually, $q_{i,1}$ is the estimate of y_d, as seen in Lemma 2, and the agents no longer require estimating the matrix P. Cooperating these two-order filters makes the use of traditional adaptive control techniques for MAS be easy, and thus the unknown time-varying control coefficients for a MAS with a directed graph can be dealt with.*

3.2. Design of the Distributed Controller

In this section, cooperating with the filter (6), the distributed adaptive controller is designed. The following error variables and change of coordinates are introduced

$$e_{i,1} = \frac{1}{k_i(t)}\left(x_{i,1} - q_{i,1} - \sigma(t)(x_{i,1}^0 - q_{i,1}^0)\right) \tag{18}$$

$$e_{i,m} = \frac{1}{k_i(t)}\left(x_{i,m} - \alpha_{i,m-1} - \sigma(t)x_{i,m}^0\right), m = 2,\dots,n \tag{19}$$

with

$$\sigma(t) = \begin{cases} \frac{1}{t_s^2}(t - t_s)^2, & t < t_s \\ 0, & t \geq t_s \end{cases} \tag{20}$$

where $x_{i,j}^0 = x_{i,j}(0), j = 1, \ldots, n$, and $q_{i,1}^0 = q_{i,1}(0)$, and t_s can be any positive constant. Let $t_s = 1$ in this paper.

Then, the intermediate control signals $\alpha_{i,m}$ and the distributed controller u_i are determined as follows

$$\alpha_{i,m} = -\lambda_{i,m}\frac{e_{i,m}}{1 - e_{i,m}^2}, \quad m = 1, \ldots, n-1 \tag{21}$$

$$u_i = -\lambda_{i,n}\frac{e_{i,n}}{1 - e_{i,n}^2} \tag{22}$$

where $\lambda_{i,m}, 1 \leq i \leq N, 1 \leq m \leq n$ are the positive design parameters. It is easy to verify that $e_{i,m}(0) = 0$ for all $1 \leq i \leq N, 1 \leq m \leq n$ and $e_{i,1}(t) = x_{i,1}(t) - q_{i,1}(t)$ for $t \geq t_s, 1 \leq i \leq N$. $k_i(t)$ are the constrained functions chosen by the designer and used as prescriptive performance functions, satisfying $0 < \underline{k} \leq k_i(t) \leq \bar{k}, \left| \dot{k}_i(t) \right| \leq k'$ with $\underline{k}, \bar{k}$ and k' being positive constants.

Remark 4. *Function $\sigma(t)$ is constructed to attenuate the influence of the initial conditions, since it makes $e_{i,m}(0) = 0$ and therefore stable results can be achieved under all initial conditions using $\sigma(t)$ for transformation (20). It should also be noted that $\sigma(t)$ of (20) is continuously differentiable and $\dot{\sigma}(t)$ does not exist in the further design of the controller, which means that the designed intermediate control signals and actual controller are smooth.*

4. Stability Analysis

In this section, we will give the main results with the designed fully distributed controller and present the stability analysis. The main results of this article are as follows.

Theorem 1. *Consider the closed-loop system consisting of N uncertain agents as (1) satisfying Assumptions 1–3, the intermediate control signals (21) and the distributed controller (22). Then, we have the following properties:*

(1)   *All the signals in the closed-loop system are globally bounded*
(2)   *Prespecified tracking performance can be guaranteed, namely,$|e_{i,1}| < 1$, for $i = 1, 2, \ldots, N$.*
(3)   *The output of each agent ultimately satisfies$|y_i - y_d| \leq k_i(t)$.*

Proof (of Theorem 1). From (18), (19) and (21), we have

$$x_{i,1} = k_i e_{i,1} + q_{i,1} + \sigma(t)(x_{i,1}^0 - q_{i,1}^0) \tag{23}$$

$$x_{i,m} = k_i e_{i,m} + \alpha_{i,m-1}(e_{i,m-1}) + \sigma(t)x_{i,m}^0, \quad m = 2, \ldots, n \tag{24}$$

It can be observed from (23) that $x_{i,1}$ is continuous function of $e_{i,1}$, $q_{i,1}$ and $\sigma(t)$, where $q_{i,1}$ and $\sigma(t)$ are bounded time-varying functions. Thus, $x_{i,1}$ can be rewritten as the form of continuous function of $e_{i,1}$ and t. Similar analysis can be made for $x_{i,m}$. Therefore, we obtain

$$\begin{aligned} \dot{e}_{i,1} &= \tfrac{1}{k_i}\left(f_{i,1}(x_{i,1}) + g_{i,1}(x_{i,1})x_{i,2} - q_{i,2} - \dot{\sigma}(t)(x_{i,1}^0 - q_{i,1}^0) - \dot{k}_i e_{i,1} + d_{i,1}(t, x_{i,1})\right) \\ &= h_{i,1}(t, e_{i,1}, e_{i,2}, \vartheta_i) \end{aligned} \tag{25}$$

$$\begin{aligned} \dot{e}_{i,m} &= \tfrac{1}{k_i}\left(f_{i,m}(\bar{x}_{i,m}) + g_{i,m}(\bar{x}_{i,m})x_{i,m+1} - \frac{\partial \alpha_{i,m-1}}{\partial e_{i,m-1}}h_{i,m-1}(t, e_{i,1}, \ldots, e_{i,m})\right. \\ &\quad \left. - \dot{\sigma}(t)x_{i,m}^0 - \dot{k}_i e_{i,m} + d_{i,m}(t, \bar{x}_{i,m})\right) \\ &= h_{i,m}(t, e_{i,1}, \ldots, e_{i,m+1}) \end{aligned} \tag{26}$$

where $e_{i,n+1} = 0$, and $h_{i,m}(\cdot)$, $m = 1, 2, \ldots, n$ are some continuous functions. Defining $e_i = [e_{i,1}, \ldots, e_{i,n}]^T$ and in view of (25) and (26), we obtain

$$\dot{e}_i = h_i(t, e_i) = \begin{bmatrix} h_{i,1}(t, e_{i,1}) \\ h_{i,2}(t, e_{i,1}, e_{i,2}) \\ \vdots \\ h_{i,n}(t, e_{i,1}, \ldots, e_{i,n}) \end{bmatrix} \tag{27}$$

Let us define the open set:

$$\Omega_e = \underbrace{(-1, 1) \times \cdots \times (-1, 1)}_{n-times} \tag{28}$$

It is easily observed that $e_i(0) \in \Omega_e, i = 1, 2, \ldots, N$. Additionally, $h_{i,m}(\cdot)$, $m = 1, 2, \ldots, n$ are continuous with respect to all its variables, owing to the fact that $y_d, \dot{y}_d, \sigma(t), q_{i,1}, k_i(t), f_{i,m}, g_{i,m}, \alpha_{i,m}$ are all continuous differentiable functions. Therefore, it follows from [32] that the conditions on $h_{i,m}(\cdot)$ ensure the existence and uniqueness of a maximal solution $\eta_i(t)$ on the time interval $[0, t_{\max})$, namely, $e_i(t) \in \Omega_e$ for $t \in [0, t_{\max})$, which implies

$$e_{i,m} \in (-1, 1), \, for \, \forall t \in [0, t_{\max}) \tag{29}$$

for $i = 1, \ldots, N$, and $m = 1, \ldots, n$.

In the following, we will prove that $t_{\max} = +\infty$ by seeking a contradiction. Suppose that $t_{\max} < +\infty$; then the related analysis is performed as follows, and all of what follows is based on $t \in [0, t_{\max})$.

Step 1: Consider the following positive definite functions

$$V_{i,1} = \frac{1}{2} \log \frac{1}{1 - e_{i,1}^2} \tag{30}$$

Let $\xi_{i,1} = \frac{1}{1 - e_{i,1}^2}$. It follows from (21), (24), (25) and (30) that the time derivative of $V_{i,1}$ is

$$\begin{aligned} \dot{V}_{i,1} &= \frac{\xi_{i,1}}{k_i}(f_{i,1}(x_{i,1}) + g_{i,1}(x_{i,1})(\alpha_{i,1} + k_i e_{i,2} + \sigma(t)x_{i,2}^0) + \\ &\quad -q_{i,2} - \dot{\sigma}(t)(x_{i,1}^0 - q_{i,1}^0) - k_i e_{i,1} + d_{i,1}(t, x_{i,1})) \\ &\leq \frac{1}{k_i} g_{i,1}(x_{i,1}) \alpha_{i,1} \xi_{i,1} + F_{i,1}(t)|\xi_{i,1}| \\ &\leq -\lambda_{i,1} E_{i,1} \xi_{i,1}^2 + F_{i,1}|\xi_{i,1}| \end{aligned} \tag{31}$$

where $F_{i,1} = \frac{1}{k_i}(|f_{i,1}(x_{i,1})| + |g_{i,1}(x_{i,1})| \left| k_i e_{i,2} + \sigma(t)x_{i,2}^0 \right| + |q_{i,2}| + \left| \dot{\sigma}(t)(x_{i,1}^0 - q_{i,1}^0) \right| + \left| k_i e_{i,1} \right| + b_{i,1}(x_{i,1}))$ and $E_{i,1} = \frac{g_{i,1}(x_{i,1})}{k_i}$. Note that $x_{i,1}$, $e_{i,1}$ and $e_{i,2}$ are bounded on Ω_e because (23) and (29), respectively. Utilizing the fact that $k_i(t), \dot{k}_i(t), \sigma(t), q_{i,1}, q_{i,2}$ are bounded and employing the extreme value theorem, owing to the continuity of $f_{i,1}(\cdot), g_{i,1}(\cdot)$ and $b_{i,1}(\cdot)$, we arrive at

$$E_{i,1} \geq c_{1,1} > 0 \tag{32}$$

$$c_{3,1} \geq F_{i,1} \geq c_{2,1} \geq 0 \tag{33}$$

where $c_{1,1}, c_{2,1}$, and $c_{3,1}$ are some unknown positive constants.

Then, substituting (32) and (33) into (31) yields

$$\dot{V}_{i,1} \leq -\lambda_{i,1} c_{1,1} \xi_{i,1}^2 + c_{3,1}|\xi_{i,1}| \tag{34}$$

From (34), it follows that $\dot{V}_{i,1}$ is negative when $|\xi_{i,1}| \leq c_{3,1}/\lambda_{i,1} c_{1,1}$ and subsequently that

$$|\xi_{i,1}| \leq \xi_{i,1}^* = \frac{c_{3,1}}{\lambda_{i,1} c_{1,1}} \tag{35}$$

which implies

$$|e_{i,1}(t)| \leq c_{4,1} = 1 - \frac{1}{\zeta_{i,1}^{*2}} < 1 \tag{36}$$

As a result, the control signal $\alpha_{i,1}$ is bounded. Moreover, invoking (24), we also can conclude the boundedness of $x_{i,2}$. Therefore, the time derivative of $\alpha_{i,1}$ is

$$\dot{\alpha}_{i,1} = -\lambda_{i,1}\dot{\zeta}_{i,1} \tag{37}$$

where

$$\left|\dot{\zeta}_{i,1}\right| \leq \frac{(1+e_{i,1}^2)}{k_i(1-e_{i,1}^2)^2}\left(|f_{i,1}(x_{i,1})| + \left|g_{i,1}(x_{i,1})(k_i e_{i,2} + \rho(t)x_{i,2}^0 + \alpha_{i,1})\right|\right) \\ + |q_{i,2}| + \left|\dot{\sigma}(t)(x_{i,1}^0 - q_{i,1}^0)\right| + \left|\dot{k}_i e_{i,1}\right| + b_{i,1}(x_{i,1})) \tag{38}$$

Noting (36) and using the same analysis as (33), it also easy to conclude the boundedness of $\dot{\zeta}_{i,1}$, and hence $\dot{\alpha}_{i,1}$.

Step j ($2 \leq j \leq n$): Consider the following positive definite functions

$$V_{i,j} = \frac{1}{2}\log\frac{1}{1 - e_{i,j}^2} \tag{39}$$

Let $\zeta_{i,j} = \frac{1}{1-e_{i,j}^2}$. In a similar fashion to that in the former step, by noting Assumption 1, it follows from (21), (24), (26) and (39) that the time derivative of $V_{i,j}$ is

$$\begin{aligned} \dot{V}_{i,j} &= \frac{\zeta_{i,j}}{k_i}(f_{i,j}(\overline{x}_{i,j}) + g_{i,j}(\overline{x}_{i,j})(\alpha_{i,j} + k_i e_{i,j+1} + \sigma(t)x_{i,j+1}^0) \\ &\quad - \dot{\alpha}_{i,j-1} - \dot{\sigma}(t)x_{i,j}^0 - \dot{k}_i e_{i,j} + d_{i,j}(t, \overline{x}_{i,j})) \\ &\leq \frac{1}{k_i}g_{i,j}(\overline{x}_{i,j})\alpha_{i,j}\zeta_{i,j} + F_{i,j}|\zeta_{i,j}| \\ &\leq -\lambda_{i,j}E_{i,j}\zeta_{i,j}^2 + F_{i,j}|\zeta_{i,j}| \end{aligned} \tag{40}$$

where $F_{i,j} = \frac{1}{k_i}(|f_{i,1}(\overline{x}_{i,j})| + |g_{i,1}(\overline{x}_{i,j})|\left|k_i e_{i,j+1} + \sigma(t)x_{i,j+1}^0\right| + |\dot{\alpha}_{i,j-1}| + \left|\dot{\sigma}(t)x_{i,j}^0\right| + \left|\dot{k}_i e_{i,j}\right| + b_{i,j}(\overline{x}_{i,j}))$ and $E_{i,j} = \frac{\pi g_{i,j}(x_{i,j})}{2k_i}$. Noting that $x_{i,m}, m = 1, 2, \ldots, j$ are bounded on Ω_e because the boundedness of $\alpha_{i,m-1}$, $e_{i,j}$ and $e_{i,j+1}$ are bounded on Ω_e in view of (29). Utilizing the fact that $k_i(t), \dot{k}_i(t)$ are bounded and employing the extreme value theorem owing to the continuity of $f_{i,j}(\cdot)$, $g_{i,j}(\cdot)$ and $b_{i,j}(\cdot)$, we arrive at

$$E_{i,j} \geq c_{1,j} > 0 \tag{41}$$

$$c_{3,j} \geq F_{i,j} \geq c_{2,j} \geq 0 \tag{42}$$

with $c_{1,j}$, $c_{2,j}$ and $c_{3,j}$ being some unknown positive constants.

Then, substituting (41) and (42) into (40) yields

$$\dot{V}_{i,j} \leq -\lambda_{i,j}c_{1,j}\zeta_{i,j}^2 + c_{3,j}|\zeta_{i,j}| \tag{43}$$

From (43), it follows that $\dot{V}_{i,j}$ is negative when $|\zeta_{i,j}| \leq c_{3,j}/\lambda_{i,j}c_{1,j}$ and subsequently that

$$|\zeta_{i,j}| \leq \zeta_{i,j}^* = \frac{c_{3,j}}{\lambda_{i,j}c_{1,j}} \tag{44}$$

which implies

$$|e_{i,j}(t)| \leq c_{4,j} = 1 - \frac{1}{\zeta_{i,j}^{*2}} < 1 \tag{45}$$

As a result, the control signal $\alpha_{i,j}$ is bounded. Moreover, we also can conclude the boundedness of $x_{i,j+1}$ by noting (24). Finally, the time derivative of $\alpha_{i,j}$ is

$$\dot{\alpha}_{i,j} = -\lambda_{i,j}\dot{\zeta}_{i,j} \tag{46}$$

where

$$\left|\dot{\zeta}_{i,j}\right| \leq \frac{(1+e_{i,j}^2)}{k_i(1-e_{i,j}^2)^2}\left(\left|f_{i,j}(\bar{x}_{i,j})\right| + \left|g_{i,j}(\bar{x}_{i,j})(k_i e_{i,j+1} + \sigma(t)x_{i,j+1}^0 + \alpha_{i,j})\right| \\ + \left|\dot{\sigma}(t)x_{i,j}^0\right| + \left|k_i e_{i,j}\right| + b_{i,j}(\bar{x}_{i,j})\right) \tag{47}$$

Noting (45) and using the same analysis as (42), it also easy to conclude the boundedness of $\dot{\zeta}_{i,j}$ and hence $\dot{\alpha}_{i,j}$.

Step n: Consider the following Lyapunov functions

$$V_{i,n} = \frac{1}{2}\log\frac{1}{1-e_{i,n}^2} \tag{48}$$

Let $\zeta_{i,n} = \frac{1}{1-e_{i,n}^2}$. Similar as the former steps, we can have

$$\dot{V}_{i,n} \leq -\lambda_{i,n}c_{1,n}\zeta_{i,n}^2 + c_{3,n}\left|\zeta_{i,n}\right| \tag{49}$$

where $c_{1,n}$ and $c_{3,n}$ are some unknown positive constants. It follows from (49) that $\dot{V}_{i,n}$ is negative when $|\zeta_{i,n}| \leq c_{3,n}/\lambda_{i,n}c_{1,n}$ and subsequently that

$$|\zeta_{i,n}| \leq \zeta_{i,n}^* = \frac{c_{3,n}}{\lambda_{i,n}c_{1,n}} \tag{50}$$

which implies

$$|e_{i,n}(t)| \leq c_{4,n} = 1 - \frac{1}{\zeta_{i,n}^{*2}} < 1 \tag{51}$$

As a result, the control signal $\alpha_{i,j}$ is bounded. Moreover, we also can conclude the boundedness of u_i. Notice that (36), (45) and (51) imply that $e_i(t) \in \Omega'_e$, for $\forall t \in [0, t_{\max}), i = 1, 2, \ldots, N$, where the set Ω'_e is nonempty and compact, defined as

$$\Omega'_e = [-c_{4,1}, c_{4,1}] \times [-c_{4,2}, c_{4,2}] \cdots \times [-c_{4,n}, c_{4,n}]$$

Owing to (36), (45) and (51) it is straightforward to verify that $\Omega'_e \subset \Omega_e$. Therefore, assuming $t_{\max} < +\infty$ dictates the existence of a time instant $t' \in [0, t_{\max})$, such that $e_i(t') \notin \Omega'_e$, which is a clear contradiction. Therefore, $t_{\max} = +\infty$. Hence, all closed-loop signals remain bounded and moreover $e_i(t) \in \Omega'_e \subset \Omega_e$, for $\forall t \geq 0$. Furthermore, from (36) we conclude that

$$|e_{i,1}(t)| \leq c_{4,1} < 1 \tag{52}$$

Then, for all $t \geq 0$. In view of Lemma 2 and (52), we have

$$\begin{aligned} \lim_{t \to +\infty}|y_i - y_d| &= \lim_{t \to +\infty}|q_{i,1} - y_d + y_i - q_{i,1}| \\ &\leq \lim_{t \to +\infty}|q_{i,1} - y_d| + \lim_{t \to +\infty}|y_i - q_{i,1}| \\ &\leq \lim_{t \to +\infty}|k_i e_{i,1}(t)| \\ &\leq k_i \end{aligned} \tag{53}$$

This completes the proof.

5. Simulation Study

Two examples will be given to demonstrate the effectiveness of the proposed distributed adaptive controller in this section, as follows.

Example 1. *Consider the following multi-agent systems*

$$\begin{cases} \dot{x}_{i,1} = g_{i,1}(\overline{x}_{i,1})x_{i,2} + f_{i,1}(t, \overline{x}_{i,1}), \\ \dot{x}_{i,2} = g_{i,2}(\overline{x}_{i,2})u_i + f_{i,2}(t, \overline{x}_{i,2}) \\ y_i = x_{i,1}, \ for \ i = 1,2,3,4 \end{cases}$$

with the system functions chosen as follows: $f_{1,1} = x_{1,1}^2$, $g_{1,1} = 1 + x_{1,1}^2$, $f_{1,2} = x_{1,1}x_{1,2}$, $g_{1,2} = 1$, $f_{2,1} = x_{2,1}^3 + 0.2\sin t$, $g_{2,1} = 1 + 0.1\cos x_{2,1}$, $f_{2,2} = x_{2,1}x_{2,2}$, $g_{2,2} = 1$, $f_{3,1} = x_{3,1}\sin x_{3,1}$, $g_{3,1} = 1$, $f_{3,2} = x_{3,1}x_{3,2} + 0.1\sin t$, $g_{3,2} = 1$, $f_{4,1} = x_{4,1} + 0.8 + 0.2\sin t$, $g_{4,1} = 1$, $f_{4,2} = x_{4,1}x_{4,2} + 0.2\cos t$, $g_{4,2} = 1$. *The communication topology for these subsystems are depicted in* Figure 1.

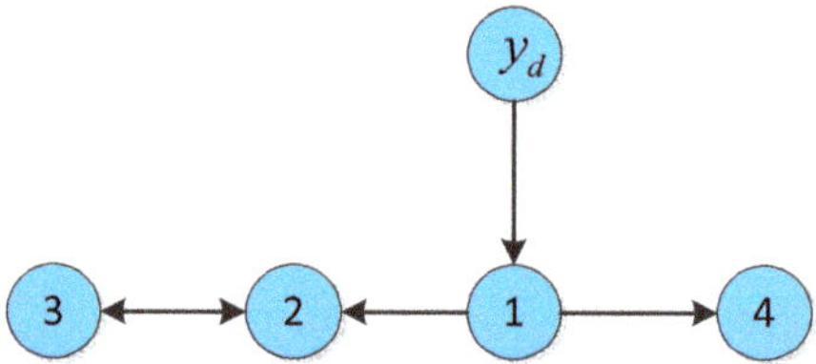

Figure 1. Communication topology for four subsystems.

The desired trajectory for the outputs of each subsystem is $y_d = \sin t$. The initial conditions for each subsystems are set as: $x_{1,1}(0) = 0.5$, $x_{2,1}(0) = -0.5$, $x_{3,1}(0) = 0$, $x_{4,1}(0) = 0.1$ and $x_{1,2}(0) = x_{2,2}(0) = x_{3,2}(0) = x_{4,2}(0) = 0$. Then, the intermediate control signals are designed and the distributed controllers are designed as follows

$$\alpha_{i,1} = -\lambda_{i,1}\frac{e_{i,1}}{1 - e_{i,1}^2}, \ i = 1,2,3,4$$

$$u_i = -\lambda_{i,2}\frac{e_{i,2}}{1 - e_{i,2}^2}, \ i = 1,2,3,4$$

where their control parameters and functions are selected as: $\lambda_{1,1} = 5$, $\lambda_{2,1} = 5$, $\lambda_{3,1} = 5$, $\lambda_{4,1} = 4$, $\lambda_{1,2} = 10$, $\lambda_{2,2} = 20$, $\lambda_{3,2} = 10$ and $\lambda_{4,2} = 10$, $k_i(t) = 3e^{-0.5t} + 0.01$ for $i = 1,2,3,4$. For the filters, the parameters are chosen as: $c_0 = 2$ and $c_1 = 6$. Then, the simulation results are reported as Figures 2–4.

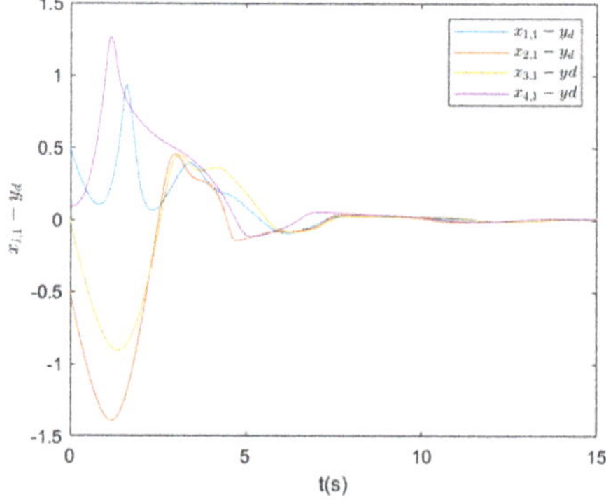

Figure 2. Tracking errors $x_{i,1} - y_d$ for $1 \leq i \leq 4$.

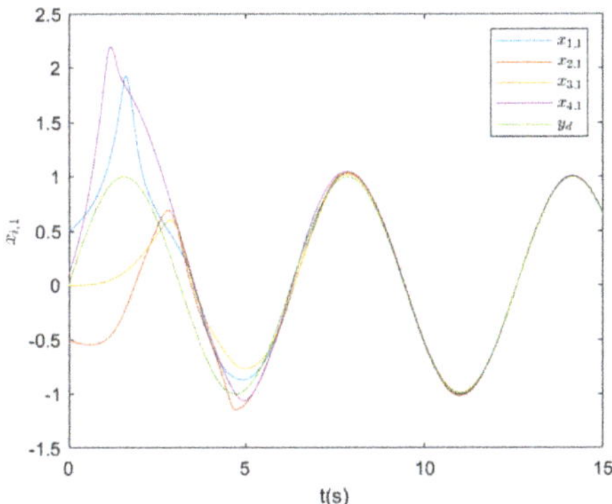

Figure 3. Outputs $x_{i,1}$ for $1 \leq i \leq 4$ and y_d.

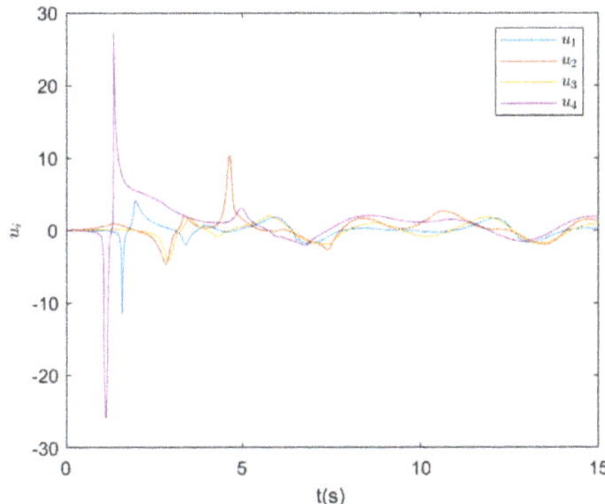

Figure 4. Distributed control inputs u_i for $1 \leq i \leq 4$.

It can be observed from Figures 2–4 that under the designed distributed controllers, the outputs of the subsystems track the desired trajectory very quick, and the tracking performance is satisfactory.

Example 2. *Consider the consensus for four high-maneuver fighters, with communication topologies as in Figure 5 and their flight control systems as follows [33].*

$$\begin{cases} \dot{X}_{i,1} = f_1(X_{i,1}, X_{i,3}) + G_1(X_{i,1})X_{i,2} \\ \dot{X}_{i,2} = f_2(X_i) + G_2 u_i \\ \dot{X}_{i,3} = f_3(X_i) \end{cases} \tag{54}$$

with

$$f_1(X_{i,1}, X_{i,3}) = \begin{bmatrix} q_i \tan \theta_i \sin \phi_i + r_i \tan \theta_i \cos \phi_i \\ p_i \beta_i + z_0 \Delta \alpha_i + (g_0/V_i)(\cos \theta_i \cos \phi_i - \cos \theta_0) \\ y_\beta \beta_i + p_i(\sin \alpha_0 + \Delta \alpha_i) + (g_0/V_i) \cos \theta_i \sin \phi_i \end{bmatrix}$$

$$G_1(X_{i,1}) = \begin{bmatrix} 1 & 0 & 0 \\ 0 & 1 & 0 \\ 0 & 0 & -\cos \alpha_0 \end{bmatrix}$$

$$f_2(X_i) = \begin{bmatrix} l_\beta \beta_i + l_p p_i + l_q q_i + l_r r_i + (l_{\beta\alpha}\beta_i + l_{r\alpha}r_i)\Delta\alpha_i - i_1 q_i r_i \\ m_\alpha \Delta\alpha_i + m_q q_i + i_2 p_i r_i - m_{\dot{\alpha}}(g_0/V_i)(\cos\theta_i \cos\phi_i - \cos\theta_0) \\ n_\beta \beta_i + n_r r_i + n_p p_i + n_{p\alpha}p_i\Delta\alpha_i - i_3 p_i q_i + n_q q_i \end{bmatrix}$$

$$G_2 = [L, M, N]^T$$

$$L = [l_{\delta_{el}}, l_{\delta_{er}}, l_{\delta_{al}}, l_{\delta_{ar}}, 0, 0, l_{\delta_r}]^T$$

$$M = [m_{\delta_{el}}, m_{\delta_{er}}, m_{\delta_{al}}, m_{\delta_{ar}}, m_{\delta_{lef}}, m_{\delta_{tef}}, m_{\delta_r}]^T$$

$$N = [n_{\delta_{el}}, n_{\delta_{er}}, n_{\delta_{al}}, n_{\delta_{ar}}, 0, 0, n_{\delta_r}]^T$$

$$f_3(X_i) = q_i \cos\phi_i - r_i \sin\phi_i$$

where $X_i = \left(X_{i,1}^T, X_{i,2}^T\right)^T = (\phi_i, \alpha_i, \beta_i, p_i, q_i, r_i, \theta_i)^T$ are the roll angle, attack angle, sideslip angle, roll angular velocity, pitching angular velocity, yaw angular velocity and pitch angle of fighter i, respectively. $y_i = X_{i,1} = [\phi_i, \alpha_i, \beta_i]^T$ $X_{i,2} = [p_i, q_i, r_i]^T$ $X_{i,3} = \theta_i$. $u_i = [\delta_{iel}, \delta_{ier}, \delta_{ial}, \delta_{iar}, \delta_{ilef}, \delta_{itef}, \delta_{ir}]^T$ are the left and right elevators, left and right ailerons, front and rear flaps, and rudder, respectively. Detailed explanations for the parameters and variables of this model can be found in [26]. Suppose that they are all flying at an altitude of 40,000 feet, at a speed of 0.6 Mach. The desired output for these fighters is $y_d = [y_{d,1}, y_{d,2}, y_{d,3}]^T = [20, 30, 0]^T$. The signal y_d is only available for fighter 1.

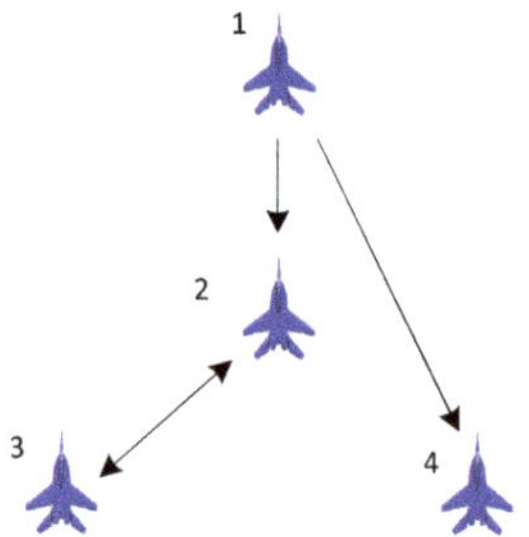

Figure 5. Communication topology for four fighters.

According to Theorem 1, we design the distributed flight controller as follows

$$\xi_{i,1} = G_1^{-1}(X_{i,1})diag\left\{-\lambda_{i,1}\frac{e_{i,\phi}}{1-e_{i,\phi}^2}, -\lambda_{i,1}\frac{e_{i,\alpha}}{1-e_{i,\alpha}^2}, -\lambda_{i,1}\frac{e_{i,\beta}}{1-e_{i,\beta}^2}\right\}$$

$$u_i = G_2^+ diag\left\{-\lambda_{i,2}\frac{e_{i,p}}{1-e_{i,p}^2}, -\lambda_{i,2}\frac{e_{i,q}}{1-e_{i,q}^2}, -\lambda_{i,2}\frac{e_{i,r}}{1-e_{i,r}^2}\right\}$$

with

$$e_{i,\phi} = \phi_i - q_{d,1}, e_{i,\alpha} = \alpha_i - q_{d,2}, e_{i,\beta} = \beta_i - q_{d,3}$$
$$[e_{i,p}, e_{i,q}, e_{i,r}]^T = [p_i, q_i, r_i]^T - \xi_{i,1}$$

where $q_{d,1}, q_{d,2}$ and $q_{d,3}$ are the signals produced by filter (6) with $y_{d,i}, i = 1, 2, 3$ being the filter inputs, respectively. $\lambda_{i,1} = 1$ and $\lambda_{i,2} = 2$ for $i = 1, 2, 3, 4$, and G_2^+ represents the pseudo-inverse for G_2.

For the purposes of comparison, we use the control method of [17] under the same conditions. Following [17], the controller for the distributed flight controller is designed as follows

$$\xi_{i,1} = G_1^{-1}(X_{i,1})diag\{-\lambda_{i,1}e_{i,\phi}, -\lambda_{i,1}e_{i,\alpha}, -\lambda_{i,1}e_{i,\beta}\}$$
$$u_i = G_2^+ diag\{-\lambda_{i,2}e_{i,p}, -\lambda_{i,2}e_{i,q}, -\lambda_{i,2}e_{i,r}\}$$

where the variables and controller parameters are the same as in our proposed methods. The simulation results are then reported in Figures 6–10. In Figure 6, the dotted curves denote the outputs of fighters under the control of the method in [17], while the solid curves denote the outputs of fighters under the control of method in this paper. It can be seen from Figure 6 that our control performance is better than [17] since the outputs of ours track the desired value more accurately. Figures 7–10 show the actions of actuators of four fighters under our method. Figure 11 show the controller performance of our method and that from [17]. In Figure 11, the blue curves denote the control efforts E_1 of the fighters with our method, while the red curves denote the control efforts E_2 of Fighters in the method from [17], where E_1 and E_2 are defined as

$$E_k = \sqrt{\delta_{iel}^2 + \delta_{ier}^2 + \delta_{ial}^2 + \delta_{iar}^2 + \delta_{ilef}^2 + \delta_{itef}^2 + \delta_{ir}^2}$$
$$k = 1,2 \ and \ i = 1,2,3,4$$

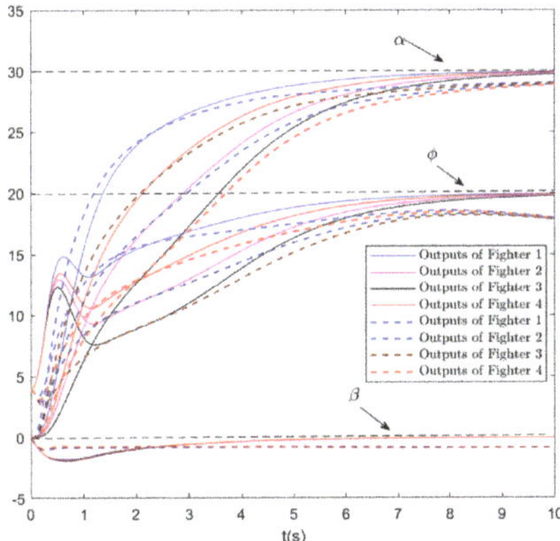

Figure 6. Output of four fighters.

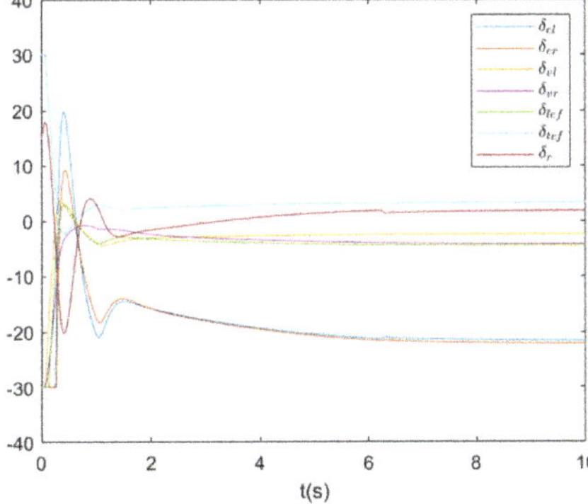

Figure 7. Actuator actions of Fighter 1.

It can be seen from Figure 11 that, initially, the control efforts of our method are greater than those in [17], and finally, there is little difference in effort between these methods, which means that the control performance of our method is better under similar control efforts.

It can be seen from these results that the consensus between the four fighters is achieved and the tracking performance is very good, while fairly good control performance is achieved.

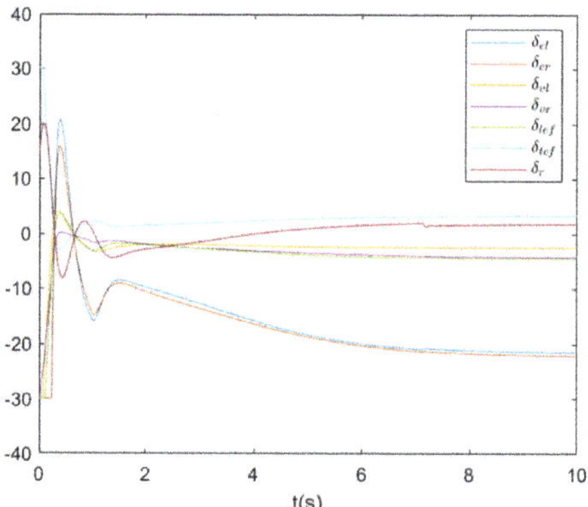

Figure 8. Actuator actions of Fighter 2.

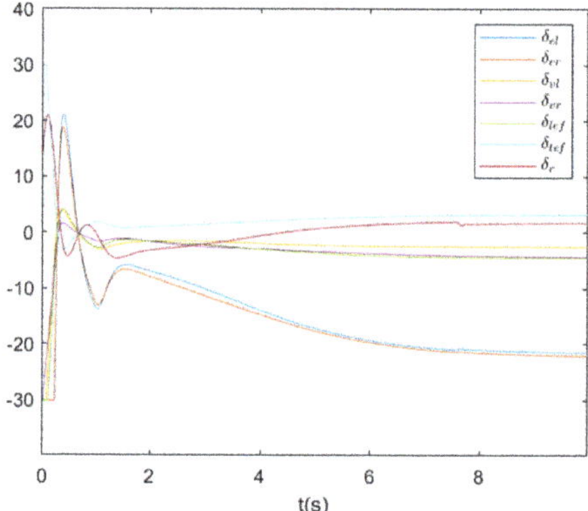

Figure 9. Actuator actions of Fighter 3.

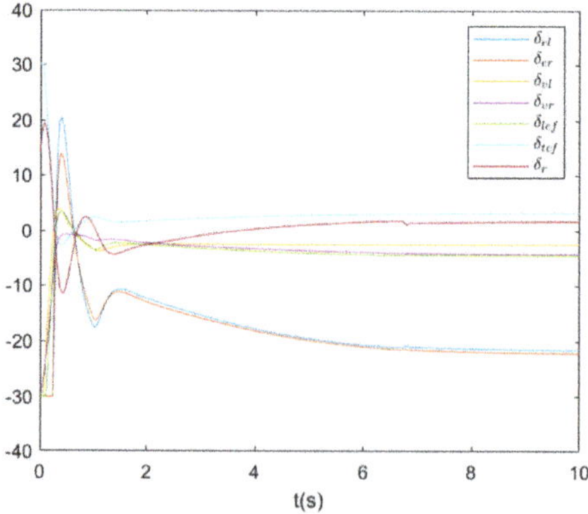

Figure 10. Actuator actions of Fighter 4.

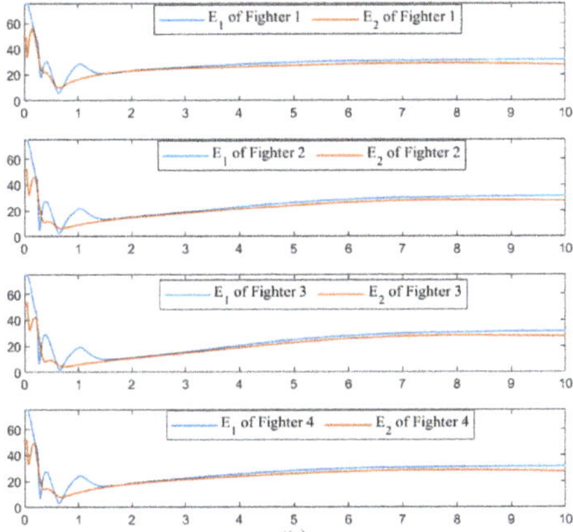

Figure 11. Control efforts.

6. Conclusions

A novel distributed consensus method was presented for a MAS with completely unknown system nonlinearities and time-varying control coefficients under a directed graph. A two-order filter for each agent was constructed, providing the desired signals and thus avoiding estimating the unknown matrix, which is related on a Laplace matrix. Combined with these filters, a global consensus method was proposed for a MAS with completely unknown system nonlinearities under a directed graph for the first time. The proposed consensus method was applied to two examples. It was shown that four high-maneuver fighters achieved angular consensus and had very good control performances using the proposed method. The two simulation results demonstrated the effectiveness of the proposed method.

Author Contributions: Funding acquisition, W.F.; Investigation, H.H.; Writing—original draft, Z.L.; Writing—review and editing, S.L. and Q.L. All authors have read and agreed to the published version of the manuscript.

Funding: This research was funded by National Natural Science Foundation of China under grant 62106284 and 62176214, and also in partly funded by Nature Science Foundation of Shaanxi Province of China under grant 2021JQ-370 and also in partly funded by Xi'an Youth Talent Promotion Plan under grant 095920201309.

Conflicts of Interest: The authors declare no conflict of interest.

References

1. Liu, Z.C.; Dong, X.M.; Xie, W.J.; Chen, Y.; Li, H.B. Adaptive fuzzy control for pure-feedback nonlinear systems with nonaffine functions being semibounded and indifferentiable. *IEEE Trans. Fuzzy Syst.* **2018**, *26*, 395–408. [CrossRef]
2. Liu, Z.C.; Dong, X.M.; Xue, J.P.; Li, H.B.; Chen, Y. Adaptive neural control for a class of pure-feedback nonlinear systems via dynamic surface technique. *IEEE Trans. Neural Netw. Learn. Syst.* **2016**, *27*, 1969–1975. [CrossRef]
3. Liu, Z.C.; Dong, X.M.; Xue, J.P.; Chen, Y. Adaptive neural control for a class of time-delay systems in the presence of backlash or dead-zone non-linearity. *IET Control Theory Appl.* **2014**, *8*, 1009–1022. [CrossRef]
4. Cai, J.P.; Wen, C.Y.; Su, H.Y.; Liu, Z.T.; Xing, L.T. Adaptive backstepping control for a class of nonlinear systems with non-triangular structural uncertainties. *IEEE Trans. Autom. Control* **2017**, *62*, 5220–5226. [CrossRef]
5. Li, Y.M.; Liu, Y.J.; Tong, S.C. Observer-based neuro-adaptive optimized control for a class of strict-feedback nonlinear systems with state constraints. *IEEE Trans. Neural Netw. Learn. Syst.* **2021**, *99*, 1–15. [CrossRef]
6. Chen, L.S. Asymmetric prescribed performance-barrier Lyapunov function for the adaptive dynamic surface control of unknown pure-feedback nonlinear switched systems with output constraints. *Int. J. Adapt. Control Signal Process.* **2018**, *32*, 1417–1439. [CrossRef]
7. Huang, J.; Wang, W.; Wen, C.; Zhou, J. Adaptive control of a class of strict-feedback time-varying nonlinear systems with unknown control coefficients. *Automatica* **2018**, *93*, 98–105. [CrossRef]

8. Li, F.Z.; Liu, Y.G. Control design with prescribed performance for nonlinear systems with unknown control directions and nonparametric uncertainties. *IEEE Trans. Autom. Control* **2018**, *62*, 3573–3580. [CrossRef]

9. Huang, X.C.; Song, Y.D.; Lai, J.F. Neuro-adaptive control with given performance specifications for strict feedback systems under full-state constraints. *IEEE Trans. Neural Netw. Learning Syst.* **2018**, *30*, 25–34. [CrossRef] [PubMed]

10. Li, Z.; Ren, W.; Liu, X.; Fu, M. Consensus of multi-agent systems with general linear and lipschitz nonlinear dynamics using distributed adaptive protocols. *IEEE Trans. Autom. Control* **2013**, *58*, 1786–1791. [CrossRef]

11. Wang, W.; Wen, C.; Huang, J. Distributed adaptive asymptotically consensus tracking control of nonlinear multi-agent systems with unknown parameters and uncertain disturbances. *Automatica* **2017**, *77*, 133–142. [CrossRef]

12. Wang, J. Distributed coordinated tracking control for a class of uncertain multiagent systems. *IEEE Trans. Autom. Control* **2017**, *62*, 3423–3429. [CrossRef]

13. Wang, W.; Wen, C.; Huang, J.; Zhou, J. Adaptive consensus of uncertain nonlinear systems with event triggered communication and intermittent actuator faults. *Automatica* **2020**, *111*, 108667. [CrossRef]

14. Huang, J.; Song, Y.; Wang, W.; Wen, C.; Li, G. Fully distributed adaptive consensus control of a class of high-order nonlinear systems with a directed topology and unknown control directions. *IEEE Trans. Cybern.* **2018**, *48*, 2349–2356. [CrossRef] [PubMed]

15. Wang, C.; Wen, C.; Guo, L. Adaptive consensus control for nonlinear multi-agent systems with unknown control directions and time-varying actuator faults. *IEEE Trans. Autom. Control* **2021**, *66*, 4222–4229. [CrossRef]

16. Bechlioulis, C.P.; Rovithakis, G.A. Decentralized robust synchronization of unknown high order nonlinear multi-agent systems with prescribed transient and steady state performance. *IEEE Trans. Autom. Control* **2017**, *62*, 123–134. [CrossRef]

17. Huang, J.; Wang, W.; Wen, C. Distributed adaptive leader–follower and leaderless consensus control of a class of strict-feedback nonlinear systems: A unified approach. *Automatica* **2020**, *118*, 109021. [CrossRef]

18. Hua, C.; You, X.; Guan, X. Leader-following consensus for a class of high-order nonlinear multi-agent systems. *Automatica* **2016**, *73*, 138–144. [CrossRef]

19. Chen, C.; Wen, C.; Liu, Z.; Xie, K.; Zhang, Y.; Chen, C.L.P. Adaptive consensus of nonlinear multi-agent systems with non-identical partially unknown control directions and bounded modelling errors. *IEEE Trans. Autom. Control* **2017**, *62*, 4654–4659. [CrossRef]

20. Huang, J.; Song, Y.D.; Wang, W.; Wen, C.; Li, G.Q. Smooth control design for adaptive leader-following consensus control of a class of high-order nonlinear systems with time-varying reference. *Automatica* **2017**, *83*, 361–367. [CrossRef]

21. Chen, J.; Li, J.; Yuan, X. Global fuzzy adaptive consensus control of unknown nonlinear multi-agent systems. *IEEE Trans. Fuzzy Syst.* **2020**, *28*, 510–522. [CrossRef]

22. Yang, H.J.; Ye, D. Adaptive fuzzy nonsingular fixed-time control for nonstrict-feedback constrained nonlinear multi-agent systems with input saturation. *IEEE Trans. Fuzzy Syst.* **2021**, *29*, 3142–3153. [CrossRef]

23. Deng, C.; Yang, G. Distributed adaptive fuzzy control for nonlinear multiagent systems under directed graphs. *IEEE Trans. Fuzzy Syst.* **2018**, *26*, 1356–1366.

24. Dong, G.; Li, H.; Ma, H. Finite-time consensus tracking neural network FTC of multi-agent systems. *IEEE Trans. Neural Net. Learning Syst.* **2021**, *32*, 653–662. [CrossRef] [PubMed]

25. Bechlioulis, C.P.; Rovithakis, G.A. A low-complexity global approximation-free control scheme with prescribed performance for unknown pure feedback systems. *Automatica* **2014**, *50*, 1217–1226. [CrossRef]

26. Zhang, J.X.; Yang, G.H. Robust adaptive fault-tolerant control for a class of unknown nonlinear systems. *IEEE Trans. Ind. Electron.* **2016**, *42*, 585–594. [CrossRef]

27. Liu, Y.J.; Tong, S.C. Barrier Lyapunov Functions-based adaptive control for a class of nonlinear pure-feedback systems with full state constraints. *Automatica* **2016**, *64*, 70–75. [CrossRef]

28. Zhang, J.X.; Yang, G.H. Adaptive prescribed performance control of nonlinear output-feedback systems with unknown control direction. *Int. J. Robust Nonlinear Control* **2018**, *28*, 4696–4712. [CrossRef]

29. Zhang, J.X.; Yang, G.H. Prescribed performance fault-tolerant control of uncertain nonlinear systems with unknown control directions. *IEEE Trans. Autom. Control* **2017**, *62*, 6529–6535. [CrossRef]

30. Sun, R.; Na, J.; Zhu, B. Robust approximation-free prescribed performance control for nonlinear systems and its application. *Int. J. Syst. Sci.* **2018**, *49*, 511–522. [CrossRef]

31. Liu, Y.H.; Li, H.Y. Adaptive asymptotic tracking using barrier functions. *Automatica* **2018**, *98*, 239–246. [CrossRef]

32. Sontag, E.D. *Mathematical Control Theory*; Springer: London, UK, 1998.

33. Oort, E.V.; Sonneveldt, L.; Chu, Q. Comparison of adaptive nonlinear control designs for an over-actuated fighter aircraft model. In Proceedings of the AIAA Guidance, Navigation & Control Conference & Exhibit, Honolulu, HI, USA, 18–21 August 2008.

applied sciences

Article

Formation Control Technology of Fixed-Wing UAV Swarm Based on Distributed Ad Hoc Network

Wenbo Suo [1], Mengyang Wang [2], Dong Zhang [2,*], Zhongjun Qu [1] and Lei Yu [3]

1. AVIC Xi'an Flight Automatic Control Research Institute, Xi'an 710076, China; suowenbo@mail.nwpu.edu.cn (W.S.); C919-zjqu@facri.com (Z.Q.)
2. School of Astronautics, Northwestern Polytechnical University, Xi'an 710072, China; wangmengyang@mail.nwpu.edu.cn
3. Science and Technology on Complex System Control and Intelligent Agent Cooperation Laboratory, Beijing 100074, China; jackzhangd@163.com

* Correspondence: zhangdong@nwpu.edu.cn; Tel.: +86-15891390256

Abstract: The formation control technology of the unmanned aerial vehicle (UAV) swarm is a current research hotspot, and formation switching and formation obstacle avoidance are vital technologies. Aiming at the problem of formation control of fixed-wing UAVs in distributed ad hoc networks, this paper proposed a route-based formation switching and obstacle avoidance method. First, the consistency theory was used to design the UAV swarm formation control protocol. According to the agreement, the self-organized UAV swarm could obtain the formation waypoint according to the current position information, and then follow the corresponding rules to design the waypoint to fly around and arrive at the formation waypoint at the same time to achieve formation switching. Secondly, the formation of the obstacle avoidance channel was obtained by combining the geometric method and an intelligent path search algorithm. Then, the UAV swarm was divided into multiple smaller formations to achieve the formation obstacle avoidance. Finally, the abnormal conditions during the flight were handled. The simulation results showed that the formation control technology based on distributed ad hoc network was reliable and straightforward, easy to implement, robust in versatility, and helpful to deal with the communication anomalies and flight anomalies with variable topology.

Keywords: fixed-wing UAV; UAV swarm formation; distributed ad hoc network; consistency theory; formation obstacle avoidance

Citation: Suo, W.; Wang, M.; Zhang, D.; Qu, Z.; Yu, L. Formation Control Technology of Fixed-Wing UAV Swarm Based on Distributed Ad Hoc Network. *Appl. Sci.* **2022**, *12*, 535. https://doi.org/10.3390/app12020535

Academic Editor: Seong-Ik Han

Received: 30 November 2021
Accepted: 30 December 2021
Published: 6 January 2022

Publisher's Note: MDPI stays neutral with regard to jurisdictional claims in published maps and institutional affiliations.

1. Introduction

The fixed-wing UAV swarm has essential application prospects and has become a current research hotspot. Completing combat missions such as coordinated reconnaissance, early warning, strike, and evaluation in the military field; and realizing disaster emergency, geological survey, and pesticide-spraying tasks in the civilian field [1–4]. Formation control is one of the key issues to achieve UAV swarm flight [5,6]. Its primary con-tents include formation maintenance, formation switching, formation obstacle avoidance, and exception handling. The distributed wireless ad hoc network [7] is the core of realizing the cluster unmanned system [8]. The formation control based on the distributed ad hoc network can better reflect the distributed, networked, and centerless characteristics of the cluster system, which is the future development trend of cluster control.

The formation control technology for UAV swarm behavior has been extensively studied. Wang [9] proposed the leader–follower method. Its basic idea is that other UAVs follow a leader UAV as followers. Luo et al. [10] and Gu et al. [11] also adopted the leader–follower method to design a control method for the leader and followers in the formation. CamPa et al. [12], based on the leader–follower method, proposed a virtual leader method. Its main idea was to treat the formation of a multi-UAV formation as a rigid virtual structure.

When the formation moved as a whole, the UAV only needed to track the movement of the fixed point corresponding to the rigid body. Li and Liu [13] designed a synchronized position tracking controller to improve the effectiveness of using a virtual structure method to maintain formation geometry. Yun and Albayrak [14] applied behavior-control methods to study the formation of multiplatform formations, such as linear formation and circular formation. Chen and Luh [15] applied behavior-control methods to achieve the purpose of object transportation. Ginlietti et al. [16] used behavior methods to define the concept of the formation geometric center. When flying in formation, each aircraft needs to maintain a prescribed distance from the geometric center, and be able to perceive the movements of other aircraft and reconstruct the formation, similar to the behavior of migratory birds in nature. Joongbo et al. [17] proposed a feedback linearization method based on consistency for multi-UAV systems to maintain a specific time-varying formation flight geometry. Glavas et al. [18] applied the consistency research strategy to study the situation when there were random communication noise and information packet loss constraints in the network, and used the polygon method based on information exchange to achieve formation control. Yasuhiro and Toru [19] studied the cooperative control problem of multi-UAV systems, and proposed a cooperative formation control strategy with collision-avoidance capability using decentralized model predictive control (MPC) and consensus-based control. Zhao et al. [20] studied the problem of formation control of multiaircraft formation with time-varying formation characteristics when there was a spanning tree in the network topology based on the consistency theory, and obtained the stability conditions of the system. Dong et al. [21] designed a distributed formation controller based on the consistency theory, proving that as long as the network topology was guaranteed to have directional strong connectivity, even if the aircraft was lost during the formation flight, the multiaircraft system could still achieve stable formation control. Seo et al. [22] designed a consistent control protocol for the situation in which the network topology had fixed connectivity, and studied the problem of cooperative formation control of multiaircraft systems forming geometric formations. Tang et al. [23] used evolutionary control theory to complete distributed collaborative control of UAV formations. He and Lu [24] proposed a decentralized design method based on a UAV distributed-formation maintenance controller, decomposing the UAV formation model into decoupled parts and associated parts using robust control methods, and improved the distributed control method of the associated system designs of the controller to control the UAV formation flying. However, most of the research content was based on the formation controller design coupled with the UAV's underlying control system. It was assumed that the UAV had a three-channel autopilot and had the ability of instantaneous response [25]. These assumptions made it difficult to apply the formation algorithm to an actual UAV swarm control. Furthermore, the research content was insufficient in the control algorithm of real flight environments such as network topology jump, communication delay, and even weak communication.

Based on the consistency theory, this paper proposes a method for formation switching and obstacle avoidance based on waypoint planning for the problem of formation control of a fixed-wing UAV swarm in a distributed ad hoc networks. The organizational structure of the paper is as follows. The first part gives a general description of the problem of a fixed-wing UAV swarm formation in a distributed ad hoc network; the second part proposes a method for switching the formation of the UAV swarm based on the consistency theory; the third part designs a UAV swarm formation obstacle avoidance algorithm; the fourth part deals with the problems of flight abnormality and communication abnormality of the UAV swarm during the flight; the fifth part simulates and verifies the formation switching of the UAV swarm, the formation obstacle avoidance, and handling of anomalies during the flight; the sixth part analyzes and discusses the results; and finally, the seventh part summarizes the article.

2. Problem Formulation

This paper focuses on two critical problems of formation switching and formation obstacle avoidance in a distributed ad hoc network for UAV swarm formation control.

Definition 1. *Formation switch in distributed ad hoc network.*

For n UAV, given an initial position $\mathbf{X}_i(0)$ of UAV_i, a UAV swarm forms a communication topology in the ad hoc network (as shown in Figure 1). Plan the waypoint $P_i = \{P_{i1}, \cdots, P_{ik}, \cdots\}$ of the UAV_i under the dynamic constraints of the maximum turning angle constraint $\beta_{\max}$ and minimum route length constraint d_s, so that the distance between UAV_i and other UAVs reaches the expected value $\Delta\mathbf{X}_{jiref}$ within the time T:

$$\left| \mathbf{X}_j(t) - \mathbf{X}_i(t) - \Delta\mathbf{X}_{jiref} \right| \to 0 \quad i = 1,2,\cdots,n \tag{1}$$

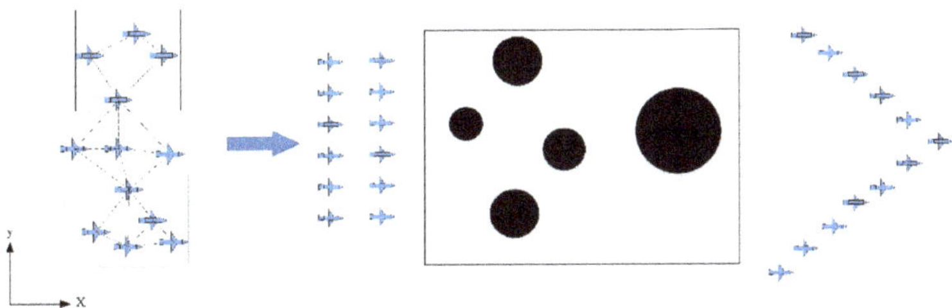

Figure 1. UAV swarm formation problem.

Definition 2. *Formation obstacle avoidance in distributed ad hoc network.*

For n UAV, given an initial position X of UAV_i, intelligently split the UAV swarm into N_c formation; the number of UAVs in the pth subformation is N_p. Plan the waypoint $P_i = \{P_{i1}, \cdots, P_{ik}, \cdots\}$ of the UAV_i under the dynamic constraints of the maximum turning angle $\beta_{\max}$ and the minimum direct flying distance d_s, so that the UAVs will not collide through the rectangular obstacle avoidance area S containing the circular obstacle O, and the distance between UAV_i the and other UAVs will eventually reach the expected value $\Delta\mathbf{X}_{jiref}$; namely, reconstructed into the required formation.

Where $n = \sum_{p=1}^{N_c} N_p$, O is a collection of obstacles, and $O = \{o_1, o_2, \cdots, o_{oN}\}$, oN is the number of obstacles, each obstacle o_m can be described as a dyadic array $< R_{om}, R_m >$; R_{om} is the center point of the mth circle; R_m is the radius of the mth circle; S is the rectangular obstacle avoidance area described as a ternary array $< S_0, L, W >$, where S_o is the center point of the rectangular obstacle avoidance area; L is the length of the rectangle; and W is the width of the rectangle.

A reasonable formation-switching control method requires that the UAV can perform online formation switching based on the neighboring UAV information, and can meet the UAV dynamic constraints and can handle communication delays and flight anomalies, eliminating track deviation. Formation obstacle avoidance requires that a UAV swarm can effectively avoid obstacles, and can form a desired formation after the avoidance is completed. The UAV studied in this paper was a highly dynamic fixed-wing UAV with a uniform speed. The dynamic constraints were required to satisfy the minimum turning radius constraints and minimum track length constraints. The research space of this paper was the two-dimensional Euclidean horizontal plane.

3. Waypoint-Based Formation-Switching Method

The UAV studied in this paper was a highly dynamic fixed-wing UAV. Due to its high speed, it was difficult for the control system to update flight parameters in real time, and the errors were relatively large. Therefore, this article mainly considered the waypoint-based formation switching, and the online design of a small number of waypoints to complete the flight process of the UAV swarm formation, without the need to participate and change the design of the aircraft control system.

3.1. Consensus-Based Design for UAV Swarm Formation Control Protocol

The algebraic graph theory was used to describe the UAV swarm system and its behavior. Assume that the ad hoc network UAV swarm system has n UAVs, and each UAV is regarded as a node, then the communication relationship is seen as an edge. An undirected graph $G = (V, E, \mathbf{A})$ represents the UAV swarm system, where $V = \{s_1, s_2, \cdots, s_n\}$ is a collection of nodes, $E = \{(s_i, s_j) \in s \times s, i \neq j\}$ is a collection of edges, and $\mathbf{A} = [a_{ij}]_{n \times n}$ represents an adjacency matrix with weights. The edge of the graph is indicated by $e_{ij} = (s_i, s_j)$. For an undirected graph, UAV_i and UAV_j can receive the information sent by each other, namely $(s_i, s_j) \in E \Leftrightarrow (s_j, s_i) \in E$. The adjacency matrix is defined as $a_{ii} = 0$, and $a_{ij} = a_{ji} \geq 0$, when $e_{ij} \in E$, $a_{ij} > 0$. The neighbor set of the node s_i is defined as $N_i = \{s_i \in V | (s_i, s_j) \in E\}$.

An undirected graph G_n was used to describe the communication topology relationship in the UAV swarm. At each moment t, the communication connection between the ad hoc network and UAV swarm forms a communication topology. For the vertex i of graph G, let $x_i(t) \in R^q$ and $u_i(t) \in R^q$ denote the state variable and state information input variable of the UAV swarm, respectively, at the time t. The classic first-order continuous-time consistency protocol [26] is:

$$\dot{x}_i(t) = u_i(t) \quad i = 1, 2, \cdots, n \tag{2}$$

$$u_i(t) = -\sum_{j=1}^{n} a_{ij}(t)\left(x_i(t) - x_j(t)\right) \tag{3}$$

For any UAV_i, the initial state is $x(0) \in R^P$, when $t \to \infty$, there is $|x_i(t) - x_j(t)| \to 0$, which is called the state of UAV swarm system reaching consensus.

In this paper, the consensus algorithm needed to be applied to the formation switching of the UAV swarm. The distance between UAVs needed to reach the expected value eventually. Therefore, to improve the classic first-order consistency protocol, the first-order consistency protocol with reference location information was proposed, and the specific form was as follows:

$$\dot{x}_i(t) = u_i(t) = \frac{\sum\limits_{j=1}^{m} \left(a_{ij}(t)[(\mathbf{X}_j(t) - \mathbf{X}_i(t)) - \mathbf{X}_{jiref}]\right)}{\sum\limits_{j=1}^{m} a_{ij}(t)} \quad i = 1, 2, \cdots, n \tag{4}$$

where $\mathbf{X}_{jiref}$ is the expected formation relative distance between UAV_j and UAV_i.

3.2. Consensus-Based Waypoints Planning

3.2.1. Formation Waypoints

The UAV speed studied in this paper was constant, and the UAV swarm formation was controlled based on the route. Therefore, the consistency control protocol in Equation (4) needed to be discretized. First, the communication time of the UAV swarm was discretized, then the position status of each UAV was updated in real time according to the difference equation, and the discrete consistency control protocol can be given by:

$$\mathbf{X}_i[k + 1] = \mathbf{X}_i[k] + \Delta\mathbf{X}_i[k] + \mathbf{D} \tag{5}$$

$$\Delta \mathbf{X}_i[k] = \frac{\sum\limits_{j=1}^{m}\left(a_{ij}[k]\left((\mathbf{X}_j[k] - \mathbf{X}_i[k]) - \mathbf{X}_{jiref}\right)\right)}{\sum\limits_{j=1}^{m} a_{ij}[k]} \tag{6}$$

where k represents a communication event, which is a formation process of the UAV swarm formation waypoint; $\mathbf{D}$ is the distance required to switch the formation, and is selected according to actual needs; $\mathbf{X}_j[k]$ and $\mathbf{X}_i[k]$ are the position status of the UAV$_j$ and UAV$_i$ at time k, also called formation waypoint; $a_{ij}[k] \in R \times R$ is the adjacency weight matrix of the communication topology, and its elements are defined as:

$$a_{ij} = \begin{cases} 1 & (v_j, v_i) \in E \\ 0 & (v_j, v_i) \notin E \end{cases} \tag{7}$$

The undirected graph in this article did not allow self-loops, so $a_{ij} = 0$.

During the flight, UAV$_i$ established a communication connection with other UAVs in the ad hoc network, forming a formation waypoint $\mathbf{X}_i[k]$ within each communication event k according to Equations (5) and (6).

According to the consistency theory, it can be proved that for any UAV$_i$ with an initial value $\mathbf{X}_i[0] \in R^P$, the time-varying communication topology union is fully connected in a formation transformation [27], namely $\sum\limits_{k=0}^{Nk} a_{ij}[k] > 0 \ (i \neq j)$, when $k \to Nk$, there is $\left|\Delta \mathbf{X}_{ji}[k] - \Delta \mathbf{X}_{jiref}[k]\right| \to 0$; that is, the relative distance between the UAV swarm reached the desired value, and the formation switch was realized. If the initial state of the communication topology $a_{ij}[0]$ was full connectivity; namely, the communication was in the normal state, the formation fly could be achieved as the expected formation when $k = 0$.

3.2.2. Flying to Formation Waypoint in Consistent Time

Once the formation waypoints were obtained, only local waypoints from the current position of the UAV$_i$ to its formation waypoint needed to be designed so that the flight time of each UAV was equal. In this section, a distributed UAV swarm flying method based on time consistency is proposed, and local waypoints are designed under dynamic constraints to achieve UAV swarm formation switching when flying at a constant speed.

(1)   Dynamic constraints

Maximum turning angle constraint: when the UAV turns according to the waypoint, the turning angle β (as shown in Figure 2) needs to be lower than the maximum turning angle $\beta_{\max}$:

$$\beta \leq \beta_{\max} \tag{8}$$

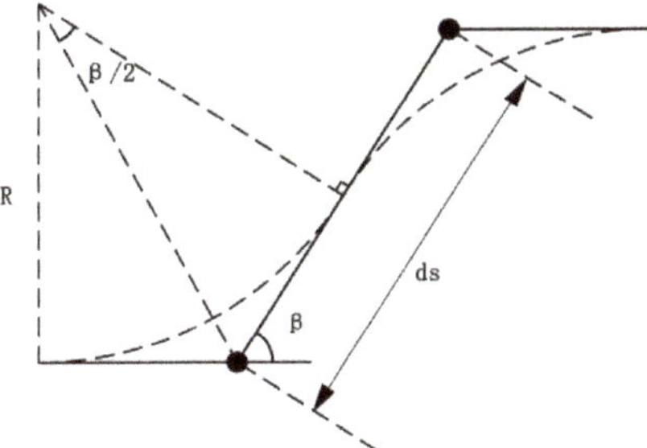

Figure 2. Dynamic constraints.

Minimum route length constraint [28]: assuming that the UAV was flying under the available overload at the turn, the turning radius R was a fixed value. As shown in Figure 2, the turning angle of the UAV in this section of the route was β. For the UAV to make two consecutive turns, the route of this section needed to be lower than the minimum track length constraint d_s:

$$d_s = 2 \times R \times \tan\frac{\beta}{2} \tag{9}$$

(2) Waypoint design

When a UAV flew to formation waypoint, we first took the UAV that was going to fly the longest path as the time base, and then planned for other UAVs to fly around in the horizontal plane so that the final time to reach the formation waypoint was consistent.

The longest path $C_{\max}$ is:

$$C_{\max} = \max(C_i) + d_y \tag{10}$$

where d_y is the vertical flight distance margin to ensure that the vertical distance meets the constraint of the minimum route length d_s. The turning angle of this method is a fixed value $\beta = 90°$, so we set $d_y = 2 \times d_s = 4 \times R$, C_i as the route distance that the UAV_i needed to fly:

$$C_i = \Delta x_i + \Delta y_i + \Delta t_i \times V \tag{11}$$

where Δx_i and Δy_i are the horizontal distance and vertical distance of UAV_i from the current position to the formation waypoint; V is the flight speed of the UAV; and Δt_i is the waiting time of UAV_i, which can be selected according to the actual project. If it was a formation switch scenario, $\Delta t_i = 0$. If it was a formation assembly scenario, Δt_i was the launch interval of UAV_i from the first UAV.

The dynamic constraints of the flight plan needed to consider the maximum turning angle constraint and the minimum route length constraint. The flight project was divided into the following sections:

As shown in Figure 3, the flight project was composed of four right-angle turning sections. The solid line is the planned route, which was sections ①, ②, ③, ④, and ⑤. The dashed line is the actual flight route considering the UAV turning process. The solid point is the UAV waypoint. To make the flying distance the same, the length of the distances of ①, ②, ③, ④, and ⑤ were designed as:

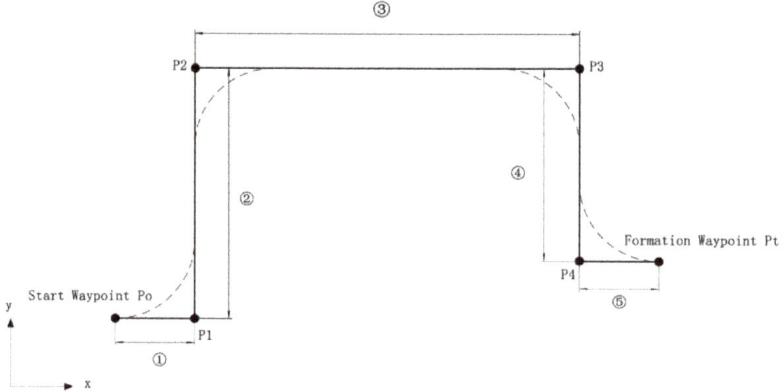

Figure 3. Dynamic constraints.

$$\begin{cases} L_{1x,i} = L_{5x,i} = r \\ L_{3x,i} = \Delta x_i - 4 \times r \\ L_{2y,i} = L_{4y,i} = \frac{C_{\max} - \Delta x_i - \Delta y_i}{2} \end{cases} \tag{12}$$

where $L_{1x,i}$, $L_{3x,i}$, and $L_{5x,i}$ are the horizontal distances of UAV$_i$ in segments ①, ③, ⑤; and $L_{2y,i}$, $L_{4y,i}$ are the vertical distances of UAV$_i$ in segments ② and ④, respectively.

The local waypoints of the UAV$_i$ could be calculated according to the starting waypoint and the distance lengths of ①, ②, ③, ④, and ⑤:

$$\begin{cases} P_{i1}(x_{i1}, y_{i1}) = P_{i0}(x_{i0} + L_{1x,i}, y_{i0}) \\ P_{i2}(x_{i2}, y_{i2}) = P_{i1}(x_{i1}, y_{i1} + L_{2y,i}) \\ P_{i3}(x_{i3}, y_{i3}) = P_{i2}(x_{i2} + L_{3x,i}, y_{i2}) \\ P_{i4}(x_{i4}, y_{i4}) = P_{i3}(x_{i3}, y_{i3} - L_{4y,i}) \end{cases} \tag{13}$$

At this point, all local route points could be obtained. The UAVs could arrive at the formation waypoints at the same time, thus forming the expected formation.

In this section, a waypoint-based distributed ad hoc network formation-switching control method, derived from the consistency theory, only needed to obtain the formation waypoint from the position information of the UAVs, and then plan the local waypoint from the current waypoint to the formation waypoint under dynamic constraints. The UAV only needed to reach the online planned waypoint under the control of its flight control system. It did not need to call the UAV's control system to track flight parameters in real time, and only required the design of four local waypoints. This method has a small amount of calculation, is practical and straightforward, and is conducive to implementation in engineering. It can also realize dynamic formation control of UAVs when some communication networks are lost and the topology structure changes.

4. Waypoint-Based Formation Obstacle Avoidance Algorithm

The traditional formation obstacle avoidance method uses an artificial potential field method for obstacle avoidance or an intelligent algorithm to search for waypoints. Still, considering the real-time nature of obstacle avoidance and engineering applications, the artificial potential field method needs to participate in the design of the control system. When facing a high-dynamics UAV, it is challenging to update flight parameters in real time, as it presents significant errors and low reliability. Using the intelligent algorithm to plan trajectories, if high precision is required, planning the trajectory of a single UAV is still too slow, and if there are many planned waypoints, it cannot meet the dynamic constraints of the UAV.

In this paper, the intelligent path search algorithm was used to search for obstacle avoidance channels through which UAVs could pass, and the UAV swarm was divided into multiple smaller formations according to the number of UAVs that could pass through the avoidance channel.

4.1. Consensus-Based Design for UAV Swarm Formation Control Protocol

Obstacle-avoidance principles include the A* algorithm and the smallest enclosing convex polygon of a set of points (SECP) decision principle.

(1)   A* algorithm

The A* algorithm [29,30] is the most effective direct search method for solving the shortest path in a static road network. We only needed to search for the formation channel between the obstacle circles, and there was no need to search for high-precision track points, so we could use the traditional A* algorithm:

$$f(o) = g(o) + h(o) \tag{14}$$

where $f(n)$ is the cost estimate from the initial state to the target state via state n; $g(n)$ is the actual cost from the initial state to state n in the state space; and $h(n)$ is the estimated cost of the best path from state to the target state. The shortest path could be determined by searching from the starting point according to the valuation function in Equation (14) to the ending point.

(2)   SECP determination method

Given a plane point set $A = \{(x_i, y_i) | x_i, y_i \in R\}$, we determined the smallest enclosing convex polygon of a set of points A_i (SECP): we connected the points two by two to form a line segment set $S = \{(A_i, A_j) | A_i, A_j \in A, i \neq j\}$. If all the other line segments were on the side of the line where a line segment S_{convex} was located, then the line segment S_{convex} where this line (e.g., the dotted line in Figure 4) lay was a side of the required convex polygon SECP.

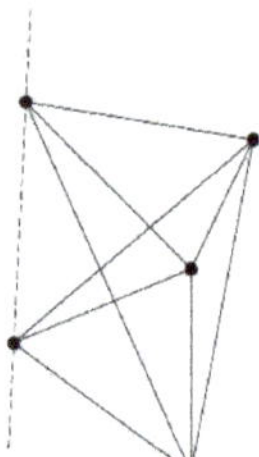

Figure 4. Determination method.

4.2. Formation Obstacle Avoidance Algorithm

To obtain multiple formation passage paths, firstly, it was necessary to determine the entry points and the exit points according to the SECP determination method; secondly, to determine the formation obstacle avoidance path in combination with the A* search method; and finally, to determine the obstacle avoidance waypoint of the UAV swarm. The algorithm flow is shown in Figure 5; the specific steps were as follows:

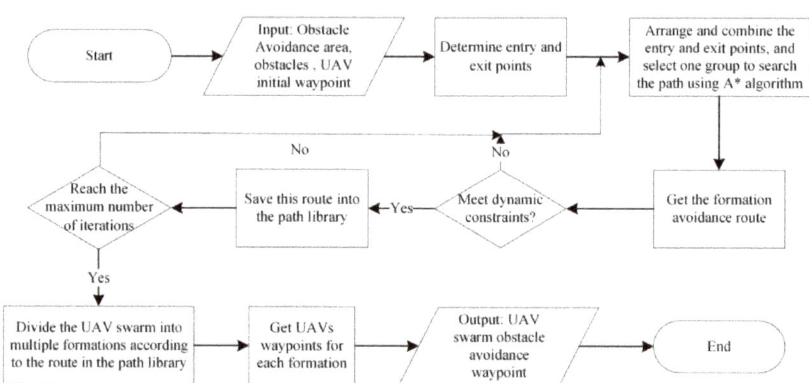

Figure 5. Principle flow chart of obstacle avoidance algorithm.

Step 1. Determine entry and exit points.

As shown in Figure 6, we connected the center points of circles two by two to get the connecting line set $CirSeg = \{CirSegment_k | k = 1, 2 \cdots, Nk\}$, and obtained the entry and exit line segments according to the SECP determination method, and further obtained the in-point set $EnPoint$ and the out-point set $ExPoint$. The specific process of the algorithm was as follows Algorithm 1:

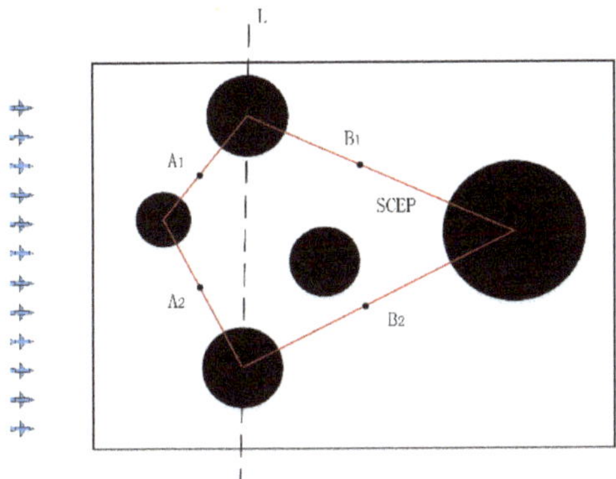

Figure 6. Determination of entry and exit points.

Algorithm 1. Get the entry points and exit points

Input: O: Obstacle collection S: Obstacle avoidance area
Output: $ExPoint = \left\{ ExitPoint_j \middle| j = 1, 2 \cdots \right\}$ $EnPoint = \{ EntryPoint_i | i = 1, 2 \cdots \}$
1: $CirSeg \leftarrow$ connect the two-point in $R_o = \{ R_{om} | m = 1, \cdots, n \}$
2: $MidLine \leftarrow$ connect the maximum and minimum of point R_{om} in y-direction
3: **for** $k = 1$ to Nk **do**
4:      $CirLine_k \leftarrow CirSegment_k$
5:      **if** $\{ CirSeg - CirSegment_k \}$ in the same side of $CirLine_k$ **then**
6:         $CirSegment_k \in SCEP$ (e.g., SCEP in Figure 6)
7:         $PathSegement \leftarrow \{ CirSegment_k - CirSegment_k \cap O \}$
8:         **if** $PathSegement$ on the left side of $MidLine$ **then**
9:            $EnPoint \leftarrow$ the middle point of $PathSegement$ (e.g., A in Figure 6)
10:        **else** $ExPoint \leftarrow$ the middle point of $PathSegement$ (e.g., B in Figure 6)
11:         **end if**
12:      **end if**
13: **end for**
14: **return** $EnPoint, ExPoint$

Step 2. Determine formation avoidance path.

As shown in Figure 7, we combined the entry and exit points set $EnPoint\ ExPoint$ from Algorithm 1 and the processed channel segment $CirSeg$ (e.g., M in Figure 7) to find the obstacle avoidance path set $AvoidPath$ with the A* method. The specific process of the algorithm was as follows Algorithm 2:

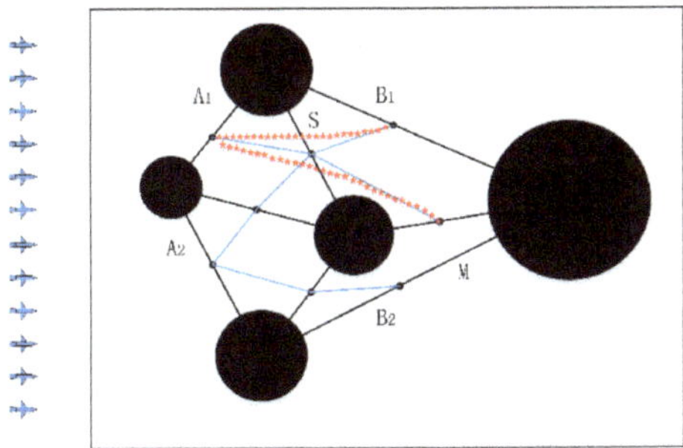

Figure 7. Determination of the formation avoidance path.

Algorithm 2. Get the avoidance path

Input: $O, S, ExPoint, EnPoint$

Output: $AvoidPath = \left\{ AvoidPath_{(i \rightarrow j),q} \,\middle|\, q = 1, 2, \cdots, Nq \right\}$

1: **for** $k = 1$ **to** Nk **do**
2:      **if** $CirSegment_k \cap O$ **then** erase $CirSegment_k$
3:      **else if** $CirSegment_k \cap \{CirSeg - CirSegment_k\}$ **then** erase $CirSegment_k$
4:      **end if**
5: **end for**
6: use O and S initialize A* map
7: **for** $i = 1$ **to** Ni **do**
8:      **for** $j = 1$ **to** Nj **do**
9:        search $Path_{i \rightarrow j}$ from $EntryPoint_i$ to $ExitPoint_j$ by A* (e.g., S in Figure 7)
10:        **for** $k = 1$ **to** Nk **do**
11:          **if** $Path_{i \rightarrow j} \cap CirSegment_k$ **then**
12:            $Pathsegement \leftarrow \{Csegment_k - Csegment_k \cap O\}$
13:            $AvoidPath_{(i \rightarrow j),q} \leftarrow$ the middle point of $Pathsegement$
14:          **end if**
15:        **end for**
16:      **end for**
17: **end for**
18: **return** $AvoidPath$ (e.g., the blue line in Figure 7)

Step 3. Determining the UAV formation obstacle avoidance waypoint.

As shown in Figure 8, we extended $AvoidPath$ to the boundary of the obstacle avoidance area (e.g., P in Figure 8), deleted the formation obstacle avoidance path that did not meet the UAV dynamic constraints, and calculated the number of UAVs that the path of the ith entry point could pass through:

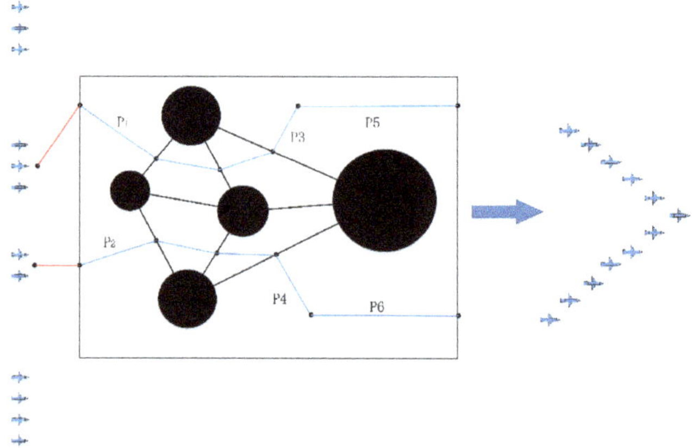

Figure 8. Determining the UAV formation obstacle avoidance waypoint.

$$n_{i \rightarrow j} = \min\left(\left\lfloor \omega \times \frac{L_{(i \rightarrow j),q} - 2 \times d_{safe}}{d} \right\rfloor + 1 \right) \quad q = 1, 2, \cdots, Nq \tag{15}$$

$$\begin{cases} n_i = \max n_{i \rightarrow j} \quad j = 1, 2, \cdots, Nj \\ jmax = \operatorname{argmax}(n_i) \\ N_{pass} = \sum\limits_{i=1}^{Ni} n_i \end{cases} \tag{16}$$

where d is the vertical interval distance of the UAV, d_{safe} is the safe distance of the UAV, $L_{(i \rightarrow j),q}$ is the length of the qth formation channel segment from the ith formation path to *j*th formation path, n_j is the number of UAV that can be passed by the *j*th formation path, ω is the scaling scale, and $0 < \omega \leq 1$ is to adjust the number of UAVs through the channels.

Then, we calculated the obstacle avoidance waypoint *PathPoint* of UAV_i based on *AvoidPath* obtained from n_i and Algorithm 2; the specific process was as follows Algorithm 3:

Algorithm 3. Get fly path point

Input: O, S, $\mathbf{X}(0)$, *AvoidPath*, v
Output: $PathPoint = \{PathPoint_{uavi} | uavi = 1, \cdots, N\}$
1: extent $AvoidPath_{i \rightarrow j}$ to S 's boundary (e.g., P in Figure 8)
2: **if** $AvoidPath_{i \rightarrow j}$ do not satisfy constraints R_{res} and d_s **then** erase
3: **end if**
4: calculate n_i by Equation (15)
5:       $CirLine_k \leftarrow CirSegment_k$
6: split UAV swarm into Ni sub-forms
7: $PathPoint \leftarrow$ calculate $PathPoint_{uavi}$ based $AvoidPath_{(i \rightarrow jmax),q}$
8: **if** $N_{pass} \leq N$ **then**
9: $PathPoint \leftarrow \{N - N_{pass}\}$ UAV fly around the obstacle area
10: **end if**
11: **for** $uavi = 1$ **to** N
12:       $T_{uavi} = |PathPoint_{uavi}|/v$
13: **end for**
14: $PathPoint \leftarrow$ use formation conversion algorithm with T_{uavi} and $PathPoint$
15: **return** $PathPoint$

---

5. Handling Exceptions

This paper dealt with two kinds of abnormal situations: flight abnormity and communication abnormalities during the formation flight of distributed ad hoc network UAV swarm. The processing methods of each category were shown in Figure 9, and the detailed processing methods were shown below.

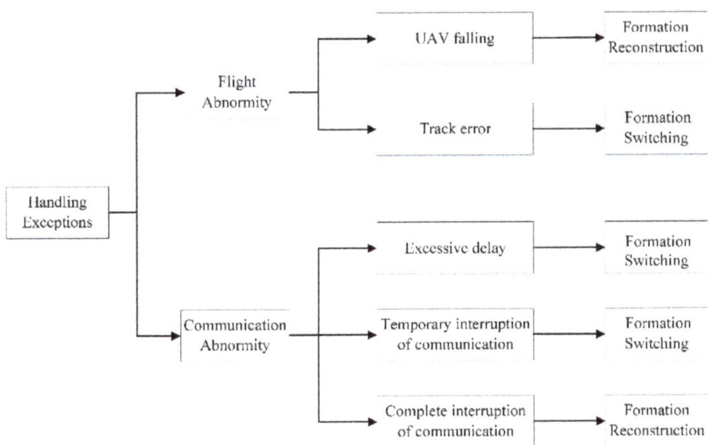

Figure 9. Handling exceptions.

5.1. Flight Abnormity

For the abnormal situation in which the tracking error was too large, the above-mentioned distributed ad hoc network online formation switch algorithm could be used for dynamic adjustment, and had the characteristics of not being related to the initial position of the UAV. In response to the falling of the UAV during the flight, the remaining

UAVs switched formation online using the formation switch algorithm; that is, formation reconstruction.

5.2. Communication Abnormalities

Communication abnormalities mainly considered three situations. First, the excessive communication delay included two cases, which were the entire UAV swarm's delay and some members' delay of the UAV swarm. Second, the communication was completely disconnected; that is, the other UAV communication links in the UAV swarm were completely disconnected and could not be restored. Third, the communication was partially interrupted; that is, the UAV communication link of other parts of the UAV swarm was temporarily disconnected, and it could be restored after some time.

If the communication of some UAVs was completely disconnected, this could be regarded as the situation of UAV formation to reconstruct the formation. If there was a delay in communication, the offset error could be eliminated according to the online formation-switching algorithm. If the communication of some UAVs was temporarily interrupted, multiple iterative formation flights could be performed based on the distributed ad hoc network online formation-switching algorithm.

6. Simulation Analysis

In this paper, 12 UAVs were used for formation assembly, switching, and formation flight simulation under abnormal flight and communication environments, and 8 UAVs were used for formation obstacle avoidance simulation. In the simulation experiment, the algorithm was programmed in the C++ language, the platform tool was Microsoft Visual Studio 2016, and the hardware environment was a PC with an inter-core i5-4210 CPU, 2.60 GHz dual-core processor, and 8 GB memory.

6.1. Formation Assembly and Formation Switching

The 12 UAVs took off in succession. After each UAV passed the assembly point A, the formation-switching algorithm began to take over the UAV, planning the trajectory of the UAV; the formation was assembled into an inverted V-shape; and then the formation-switching algorithm was enabled again to change the swarm into a V-shaped formation, as shown in Figure 10. The simulation parameters that had to be set are shown in Table 1. The simulation results of formation assembly and formation switching in the horizontal plane are shown in Figures 11 and 12.

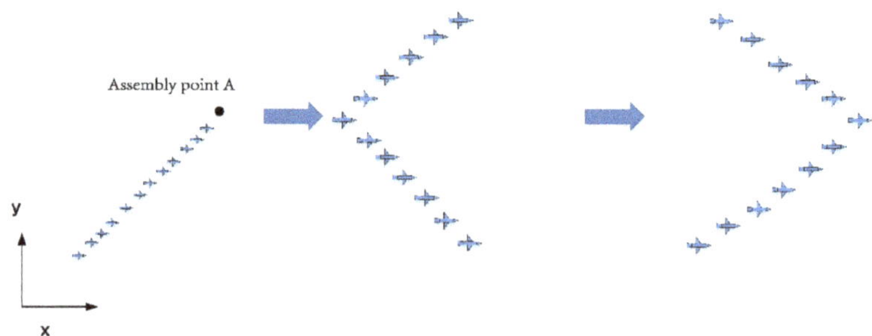

Figure 10. UAV swarm formation assembly and formation switching.

Table 1. Simulation parameters.

UAV Swarm Attributes	Parameter Value
Number of UAVs n	12
Launch interval	5 s
UAV speed V	30 m/s
Maximum turning angle constraint β_{max}	90°
Turning radius R	300 m
Minimum track length constraint d_s	600 m
Assembly point A	(0 m, 2000 m)
Communication topology	Fully Connected
Assembled formation	Inverted V-Shape
Switch formation	V-Shape
Formation interval in x direction	100 m
Half vertex angle of V-shaped	45°
Formation-switching distance D	Adaptive (met the minimum switching distance)

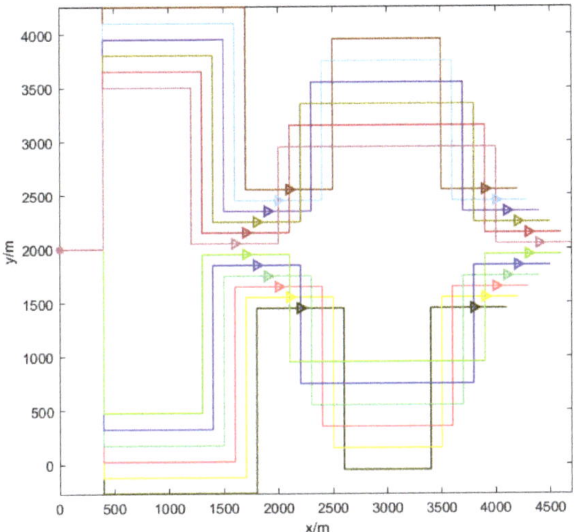

Figure 11. Formation assembly and formation switching.

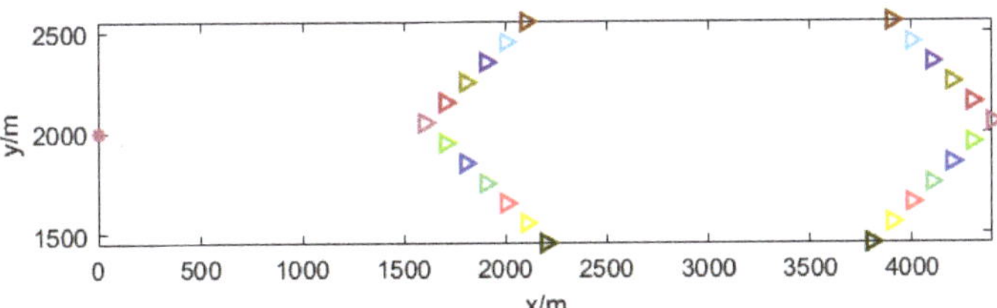

Figure 12. Assembly formation and switch formation.

The assembly formation was V-shaped, the formation switching was changed from V-shaped to column formation, the vertical formation interval in the y-direction was 100 m, and other parameters remain unchanged. The simulation is shown in Figures 13 and 14.

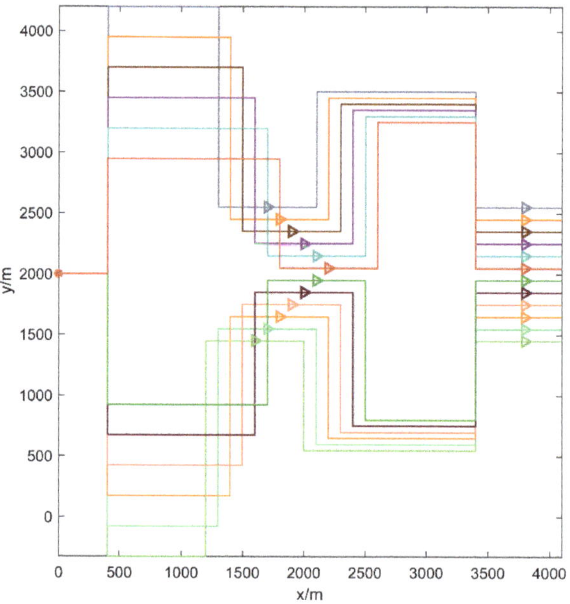

Figure 13. Formation assembly and formation switching—scene 2.

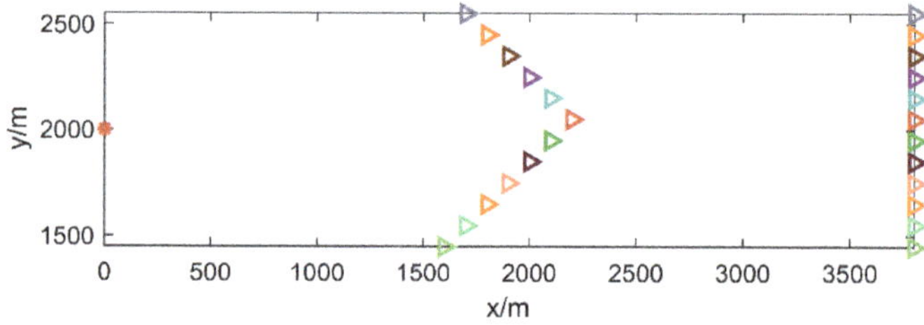

Figure 14. Assembly formation and switch formation—scene 2.

It can be seen in Figures 12 and 14 that when the communication topology of UAV swarm was fully connected, formation aggregation and formation switching could make the distance between each UAV meet the desired requirements after a formation interval correction; namely, the communication event was $k = 0$. If the communication topology was partially interrupted, the formation interval had to be corrected again. This method could simply and quickly converge to the final expected formation interval value. It can be seen in Figures 11 and 13 that this formation method could quickly plan the route to the formation waypoints under the constraints of the UAV dynamics, and has a high engineering application value.

6.2. Formation Obstacle Avoidance

The obstacle avoidance area was rectangular, and the relevant parameters are shown in Table 2. Eight UAVs formed a column formation to enter the obstacle avoidance area. The obstacle information is shown in Table 3. The initial position information of UAVs is shown in Table 4. The maximum turning angle constraint was 90°, and the minimum direct flight distance constraint was calculated according to Equation (9). The UAV swarm passed through the obstacle and was reconstructed into a V-shape. Other parameters of UAV swarm are shown in Table 1. The horizontal plane simulation results are shown in Figure 15.

Table 2. Obstacle avoidance area parameters.

Obstacle Avoidance Area Properties	Parameter Value
Rectangle center point	(7000 m, 5000 m)
Rectangular area length	10,000 m
Rectangular area width	10,000 m

Table 3. Obstacle parameters.

Obstacle Index	X/m	Y/m	Radius/m
1	3000	7000	500
2	4000	3000	500
3	6500	5000	500
4	11,000	3000	900
5	11,000	6000	600

Table 4. Initial position of UAV.

UAV Index	X/m	Y/m
1	1000	3000
2	1000	6500
3	1000	3500
4	1000	6000
5	1000	4000
6	1000	5500
7	1000	4500
8	1000	5000

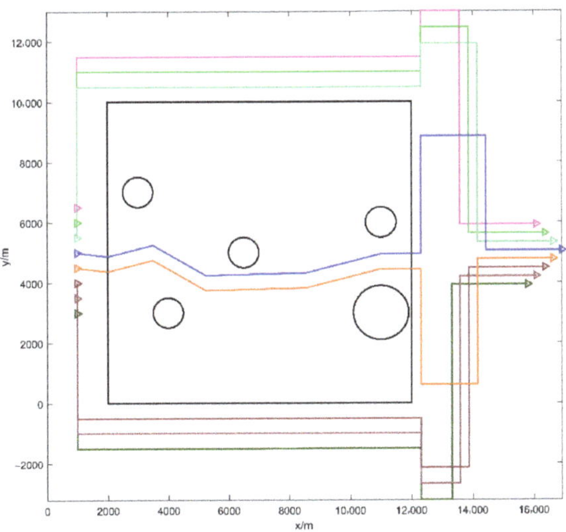

Figure 15. UAV formation obstacle avoidance.

We reset obstacles for simulation, and the obstacle information is shown in Table 5. Other parameters remain unchanged; the horizontal plane simulation results are shown in Figure 16.

Table 5. Obstacle parameters—scene 2.

Obstacle Index	X/m	Y/m	Radius/m
1	3000	5000	700
2	6000	14,000	1000
3	6500	5000	700
4	7000	8500	1300
5	11,000	4000	900

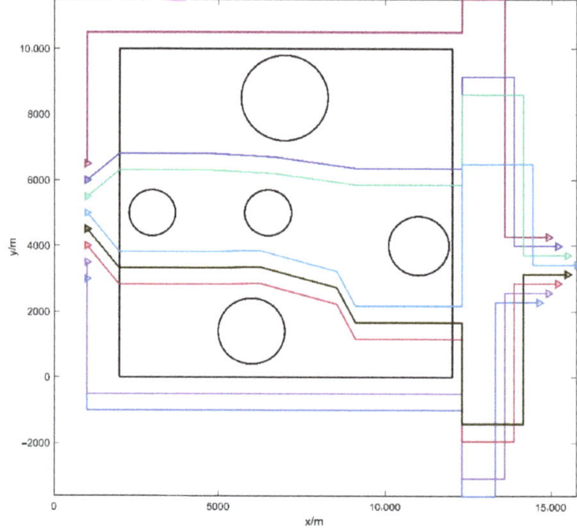

Figure 16. UAV formation obstacle avoidance—scene 2.

As can be seen in Figures 15 and 16, the UAV swarm could be intelligently divided into multiple smaller formations to pass obstacles or fly around to avoid obstacles. Finally, it could be reconstructed into the expected formation. The simulations showed that the algorithm could quickly and flexibly plan the cooperative obstacle avoidance path and had the ability of online formation to avoid obstacles.

6.3. Handle Exceptions

6.3.1. Flight Abnormity

Suppose the number of UAV swarm was 12, and the assembly formation was an inverted V-shaped formation. After the formation assembly, four UAVs were randomly dropped or lost. Then, the remaining eight UAVs were reconstructed into a column formation. Other simulation parameters were still as shown in Table 1, and the simulation results are shown in Figures 17 and 18:

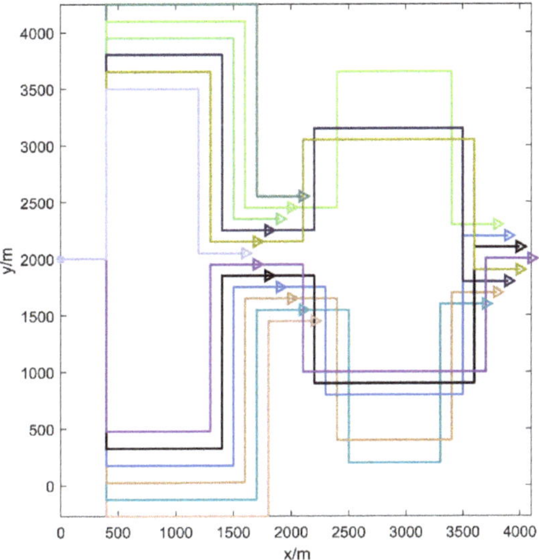

Figure 17. Formation reconstruction.

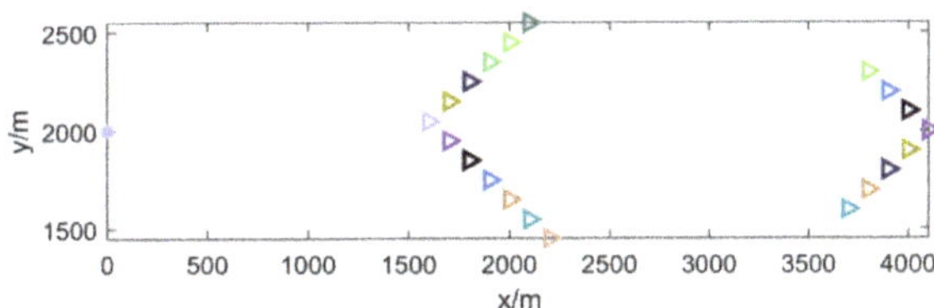

Figure 18. Formation reconstruction process.

It can be seen in Figures 17 and 18 that after the UAV flight abnormally fell, then the required formation could be reconstructed according to this plan.

6.3.2. Communication Abnormity

In actual flight, the communication topology may be temporarily interrupted. At this time, a communication event, $k = 0$, is insufficient to meet the expected formation, and formation interval distance correction is required again.

In this simulation, a formation switch was performed from an inverted V-shape to a V-shape. It was assumed that in the initial communication event, namely $k = 0$, the communication failure rate reached 18%, and in the second communication event, namely $k = 1$, the communication failure rate reached 9%. Other parameters were as shown in Table 1, and the results are shown in Figures 19 and 20.

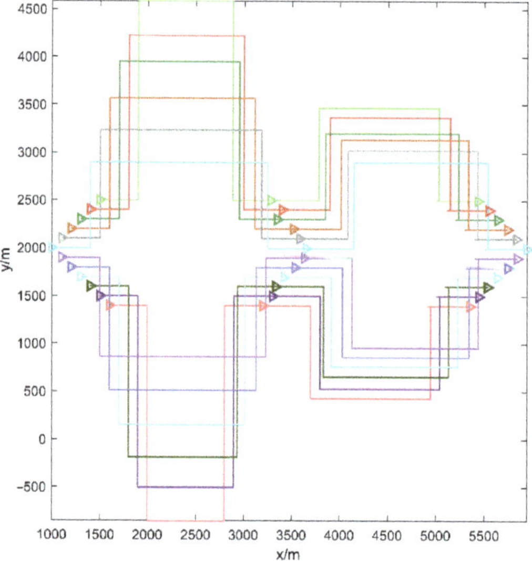

Figure 19. Formation switch under abnormal communication.

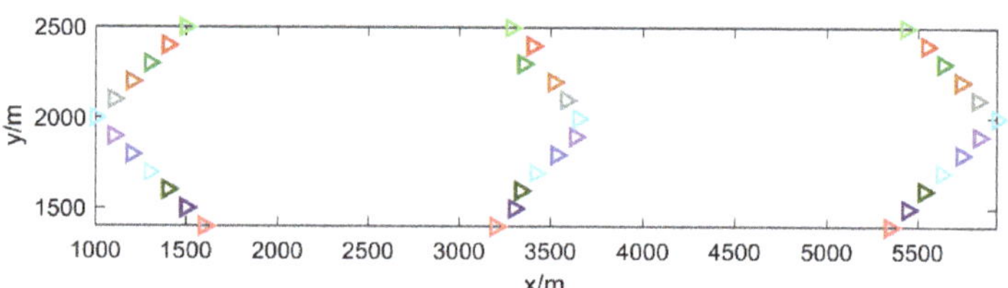

Figure 20. Formation change process under abnormal communication.

It can be seen in Figures 19 and 20 that in the case in which some UAVs and other UAVs were lost in the communication network, the first formation interval distance correction could not obtain the desired formation. In the second correction, the communication network still had local faults, but eventually formed the expected formation. So, the formation-switching algorithm was effective in dealing with communication abnormity.

6.4. Simulation Comparison

6.4.1. Formation-Switching Method

The formation-switching control method proposed in this paper mainly dealt with the communication abnormalities of the UAVs during the flight. Therefore, the classic consistency control method [31], which could handle changes in the communication topology, was selected to compare to illustrate the effectiveness of the algorithm.

We selected six UAVs for simulation, and used the simulation scenario in Section 6.3.2 to set the consistency control step to meet the minimum algorithm stability requirement, which was 0.2 s. The simulation of the classical consistency control method is shown in Figure 21, and the method proposed in this paper is shown in Figure 22.

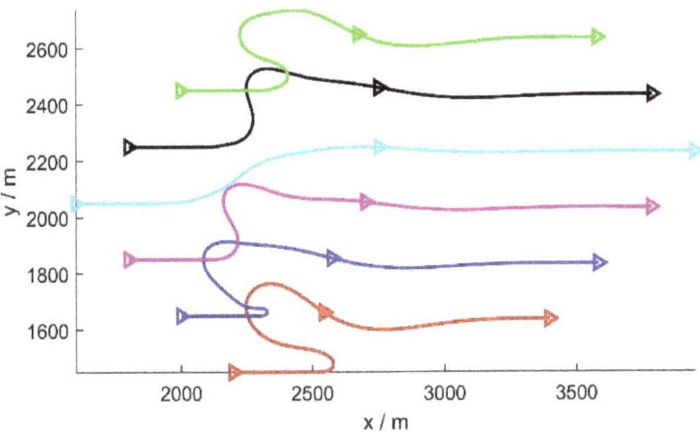

Figure 21. Formation switching of the classical consistency control method.

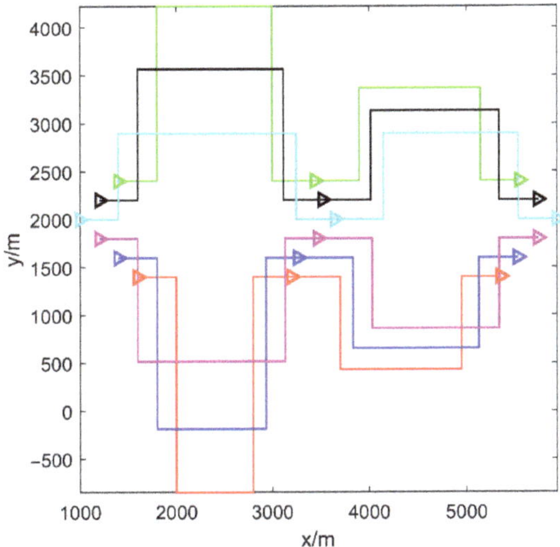

Figure 22. Formation switching.

The algorithm performance during the simulation process was recorded, as shown in Table 6.

Table 6. Simulation comparison.

Method	Classical Consistency Control Method	Proposed Method
Number of communication requests	200 times	2 times
Average solution time per communication	125 ms	53 ms
Total running time of the algorithm	1521 ms	131 ms

The simulation results showed that when the step length was 0.5 s, in the case of abnormal communication, the classic consistent formation control method needed to communicate with neighboring UAVs 200 times before the formation could be changed from an inverted V to a V-shape. The route-based formation-switching method proposed in this paper only needed two communications to obtain the required formation waypoints, which greatly reduced the pressure on the airborne communication system.

In addition, the classic consensus algorithm obtained the coordinated route of the UAV by controlling the deflection angle of the track. The average solution time per communication took a longer time, and required high control system performance, which could not guarantee the real-time performance of online formation flying. The solution proposed in this paper was time-consuming, and only required the UAV to perform direct flight and turning maneuvers. It was easy to control during the flight and easy to implement in engineering. However, the disadvantage was that the planned path distance was relatively long.

6.4.2. Formation Obstacle Avoidance Method

The sparse A * algorithm [30] was selected and compared with the formation obstacle avoidance algorithm proposed in this paper to illustrate the effectiveness of the algorithm. The simulation scenario used scenario 2 as given in Section 6.2. The sparse A* simulation results are shown in Figure 23.

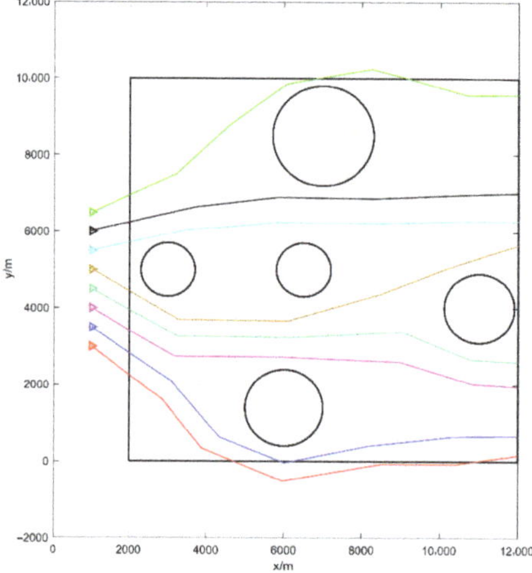

Figure 23. The sparse A * algorithm for formation obstacle avoidance.

In order to analyze the complexity of the two algorithms, the number of UAVs in the formation was increased successively, and the two algorithms were used to simulate the formation obstacle avoidance simulation. The relationship between the algorithm running time and the number of drones was obtained as shown in Figure 24.

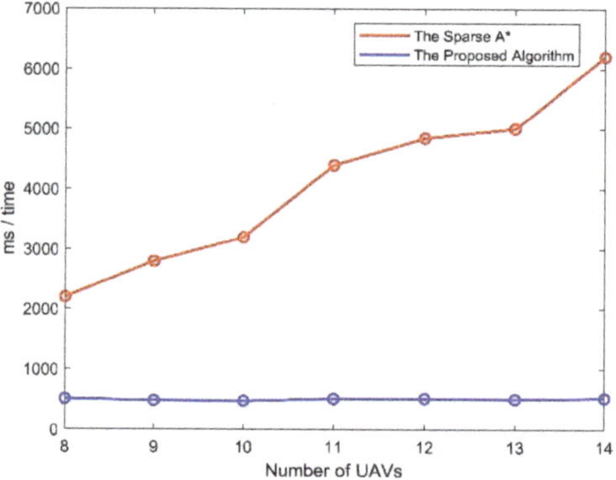

Figure 24. The algorithm running time.

We concluded from Figures 16 and 24 that the sparse A* algorithm was basically the same as the obstacle avoidance path planned by the algorithm in this paper, but the running time of the sparse A* algorithm was much longer than that of the algorithm in this paper, and it doubled with the increase in the number of UAV formations. However, the obstacle avoidance algorithm proposed in this paper based on geometry had a shorter running time and did not change significantly with the increase in the number of UAVs. It could meet online real-time trajectory planning, and has strong engineering applicability.

7. Conclusions

This paper proposed a method of fixed-wing UAV swarm formation control based on distributed ad hoc networks. This method included the formation switching of UAVs, formation obstacle avoidance, and handling of abnormal conditions during flight.

Simulation results showed that in formation switching, the ad hoc network, high-dynamics, fixed-wing UAV could form formation switching only by position information. The method was not sensitive to the initial position information of the UAV, which could eliminate errors during flight, and handle temporarily interrupted communication topologies and UAV drop, as well as other abnormal flight situations. In formation obstacle avoidance, the UAVs could be clustered into multiple subformations to pass through the obstacle avoidance area, and could be reconstructed as required formation. The formation technology was designed based on the waypoints, which was versatile, simple, and reliable, and is easy to realize in engineering.

Compared with the classic consistent formation control algorithm and obstacle avoidance algorithm, it was shown that the formation technology method proposed in this paper had lower complexity and higher timeliness, and is suitable for online formation flying of highly dynamic UAVs in an ad hoc network. However, the route planned by the method did not consider optimality, and it was only suitable for high-speed and constant-speed formation missions. For obstacle avoidance algorithms, the obstacle model is too simple, and there may be scenarios for obstacle avoidance that are not covered by the algorithm. Next, we will consider the optimal route planning problem and extend the algorithm to three-dimensional flight scenes to increase its practicability.

8. Patents

Suo, W.B., Zhang, D., and Wang, M.Y. "A distributed unmanned aerial vehicle flying around formation method based on time consistency", C.N. Patent, 202010226440.4, issued 10 July 2020.

Zhang, D., Suo, W.B., and Wang, M.Y. "A distributed unmanned aerial vehicle dynamic formation switching method based on waypoints", C.N. Patent, 202010226439.1, issued 10 July 2020.

Author Contributions: Conceptualization, W.S. and D.Z.; methodology, W.S.; software, W.S.; validation, W.S.; formal analysis, W.S.; investigation, W.S. and D.Z.; resources, D.Z. and Z.Q.; data curation, W.S., M.W. and L.Y.; writing—original draft preparation, W.S.; writing—review and editing, W.S.; visualization, W.S.; supervision, D.Z.; project administration, Z.Q.; funding acquisition, Z.Q. All authors have read and agreed to the published version of the manuscript.

Funding: This research received no external funding.

Institutional Review Board Statement: Not applicable.

Informed Consent Statement: Not applicable.

Data Availability Statement: The data presented in this study are available upon request from the corresponding author.

Conflicts of Interest: The authors declare no conflict of interest.

References

1. International Civil Aviation Organization. *Unmanned Aircraft Systems (UAS)*; International Civil Aviation Organization: Montreal, QC, Canada, 2011; p. 8.
2. Duan, H.B.; Shen, Y.K.; Wang, Y.; Luo, D.L.; Niu, Y.F. Review of technological hot spots of unmanned aerial vehicle in 2018. *Sci. Technol. Rev.* **2019**, *37*, 82.
3. Plett, G.L.; Lima, P.D.; Pack, D.J. Target Localization using Multiple UAVs with Sensor Fusion via Sigma-Point Kalman Filtering. In Proceedings of the AIAA Infotech Aerospace Conference and Exhibit, Rohnert Park, CA, USA, 7–10 May 2007. [CrossRef]
4. Toussaint, G.J.; Lima, P.D.; Pack, D.J. Localizing RF Targets with Cooperative Unmanned Aerial Vehicles. In Proceedings of the American Control Conference, New York, NY, USA, 9–13 July 2007; pp. 5928–5933. [CrossRef]
5. Duan, H.B.; Qiu, H.X.; Chen, L. Prospects on unmanned aerial vehicle autonomous swarm technology. *Sci. Technol. Rev.* **2018**, *36*, 90.
6. Zong, Q.; Wang, D.D.; Shao, S.K.; Zhang, B.; Han, Y. Research status and development of multi UAV coordinated formation flight control. *J. Harbin Inst. Technol.* **2017**, *49*, 1–14. [CrossRef]
7. Perkins, C.E.; Belding-Royer, E.M. Ad-hoc On-Demand Distance Vector Routing. In Proceedings of the IEEE Workshop on Mobile Computing Systems & Applications, New Orleans, LA, USA, 25–26 February 1999; pp. 90–100. [CrossRef]
8. Bekmezci, I.; Sahingoz, O. Flying Ad-Hoc Networks (FANETs): A survey. *Ad Hoc Netw.* **2013**, *11*, 1254–1270. [CrossRef]
9. Wang, P.K.C. Navigation strategies for multiple autonomous mobile robots moving in formation. *J. Robot. Syst.* **1991**, *8*, 177–195. [CrossRef]
10. Luo, D.; Zhou, T.; Wu, S. Obstacle Avoidance and Formation Regrouping Strategy and Control for UAV Formation Flight. In Proceedings of the 10th IEEE International Conference on Control and Automation, Hangzhou, China, 12–14 June 2013; pp. 1921–1926. [CrossRef]
11. Gu, Y.; Seanor, B.; Campa, G.; Napolitano, M.R.; Rowe, L.; Gururajan, S.; Wan, S. Design and Flight Testing Evaluation of Formation Control Laws. *IEEE Trans. Control Syst. Technol.* **2006**, *14*, 105–1112. [CrossRef]
12. CamPa, G.; Napolitano, M.R.; Brad, S.; Perhinschi, M.G. Design of Control Laws for Maneuvered Formation Flight. In Proceedings of the 2004 American Control Conference, Boston, MA, USA, 30 June–2 July 2004; pp. 2344–2349. [CrossRef]
13. Li, N.H.M.; Liu, H.H.T. Formation UAV Flight Control using Virtual Structure and Motion Synchronization. In Proceedings of the American Control Conference, Seattle, WA, USA, 11–13 June 2008; pp. 11–13. [CrossRef]
14. Yun, X.; Alptekin, G.; Albayrak, O. Line and Circle Formation of Distributed Physical Mobile Robots. *J. Robot. Syst.* **1997**, *14*, 63–76. [CrossRef]
15. Chen, Q.; Luh, J.Y.S. Coordination and Control of a Group of Small Mobile Robots. In Proceedings of the IEEE International Conference on Robotics and Automation, San Diego, CA, USA, 8–13 May 1994; pp. 2315–2320. [CrossRef]
16. Ginlietti, F.; Pollini, L.; Innoeenti, M. Formation Flight Control: A Behavioral Approach. In Proceedings of the AIAA Guidance, Navigation and Control Conferenee, Montreal, QC, Canada, 6–9 August 2001. [CrossRef]
17. Joongbo, S.; Chaeik, A.; Youdan, K. Controller Design for UAV Formation Flight Using Consensus Based Decentralized Approach. In Proceedings of the AIAA Aerospace Conference, Seattle, WA, USA, 6–9 April 2009; p. 1826. [CrossRef]
18. Glavas, K.L.S.; Chaves, M.; Day, R.; Nag, P.; Williams, A.; Zhang, W. Vehicle Networks: Achieving Regular Formation. In Proceedings of the IEEE American Control Conference, Denver, CO, USA, 4 July 2003; pp. 4095–4100. [CrossRef]
19. Yasuhiro, K.; Toru, N. Consensus-based cooperative formation control with collision avoidance for a multi-UAV system. In Proceedings of the IEEE American Control Conference, Portland, OR, USA, 4–6 June 2014; pp. 4–6. [CrossRef]
20. Zhao, S.; Lin, F.; Peng, K.; Chen, B.M.; Lee, T.H. Finite-time Stabilization of Circular Formations using Bearing-only Measurements. In Proceedings of the 10th IEEE International Conference on Control and Automation, Hangzhou, China, 12–14 June 2013; pp. 840–845. [CrossRef]
21. Dong, X.W.; Yu, B.C.; Shi, Z.Y.; Zhong, Y.S. Time-varying formation control for unmanned aerial vehicles: Theories and applications. *IEEE Trans. Control Syst. Technol.* **2014**, *23*, 340–348. [CrossRef]
22. Seo, J.; Kim, Y.; Kim, S.; Tsourdos, A. Consensus-based reconfigurable controller design for unmanned aerial vehicle formation flight. *Proc. Inst. Mech. Eng. Part G J. Aerosp. Eng.* **2012**, *226*, 817–829. [CrossRef]
23. Tang, Y.; Gao, H.J.; Jürgen, K.; Fang, J. Evolutionary Pinning Control and Its Application in UAV Coordination. *IEEE Trans. Ind. Inform.* **2012**, *8*, 828–838. [CrossRef]
24. He, Z.; Lu, Y.P. Decentralized Design Method of UAV Formation Controllers for Formation Keeping. *Acta Aeronaut. Astronaut. Sin.* **2008**, *29*, 55–60. [CrossRef]
25. Bai, C.; Duan, H.B.; Li, C.; Zhang, Y.P. Dynamic multi-UAVs formation reconfiguration based on hybrid diversity-PSO and time optimal control. In Proceedings of the IEEE Intelligent Vehicles Symposium, Xi'an, China, 5 July 2009; pp. 775–779. [CrossRef]
26. Ren, W.; Randal, W.B. *Distributed Consensus in Multi-Vehicle Cooperative Control*; Springer: London, UK, 2008; p. 21.
27. Xavier, D.; Konagaya, A. Circle Formation for Oblivious Anonymous Mobile Robots with no Common Sense of Orientation. In Proceedings of the 2002 Workshop on Principles of Mobile Computing, Toulouse, France, 30–31 October 2002; pp. 30–31. [CrossRef]
28. Yao, D.D.; Wang, X.F.; Tian, Z. A Cooperative Path Planning with Constraints of Impact Time and Impact Angle. *J. Proj. Rock Miss. Guid.* **2019**, *39*, 111–114.

29. Hart, P.E.; Nilsson, N.J.; Raphael, B. A Formal Basis for the Heuristic Determination of Minimum Cost Paths. *IEEE Trans. Syst. Sci. Cybern.* **1968**, *4*, 100–107. [CrossRef]
30. Li, J.; Sun, X.X. A Route Planning's Method for Unmanned Aerial Vehicles Based on Improved A-Star Algorithm. *ACTA Armamentarii* **2008**, *29*, 788–792. [CrossRef]
31. Xue, R.B. Research on Distributed Cooperative Formation Flight Control Technology for Multi-UAV. Ph.D. Thesis, Beijing Institute of Technology, Beijing, China, June 2016.

Article

Distributed Grouping Cooperative Dynamic Task Assignment Method of UAV Swarm

Boyu Qin [1,2], **Dong Zhang** [1,2,*], **Shuo Tang** [1,2] **and Mengyang Wang** [1,2]

[1]   School of Astronautics, Northwestern Polytechnical University, Xi'an 710072, China;
byqin@mail.nwpu.edu.cn (B.Q.); stang@nwpu.edu.cn (S.T.); wangmengyang@mail.nwpu.edu.cn (M.W.)

[2]   Shannxi Aerospace Flight Vehicle Design Key Laboratory, Xi'an 710072, China

*   Correspondence: zhangdong@nwpu.edu.cn; Tel.: +86-15891390256

Abstract: Aiming at the problem of UAV swarms with distributed subsets performing cooperative reconnaissance-and-attack tasks on multi-targets in complex and uncertain combat scenarios, a distributed grouping cooperative dynamic task assignment method is proposed based on extended contract network protocol. The dynamic task assignment model for the UAV swarm with the topology of distributed subsets is established considering multiple constraints such as task cooperation, performing sequence, dynamic environment, communication topology, payload model, and UAV capability. According to the characteristics of multi-participants and multi-tasks in the process of UAV swarm executing tasks, the determination mechanism on cooperators and the selection mechanism of sequential tasks are proposed, and then the contract network protocol is extended. On the basis of the above, an event-triggered task assignment strategy for dynamic tasks is designed. The simulated results show that the proposed method can achieve the cooperative dynamic assignment of the UAV swarm to perform reconnaissance-and-attack tasks to multi-targets in complex and uncertain combat scenarios, improve the adaptiveness of the swarm under the sudden circumstance, and realize the optimization for task execution efficiency of the UAV swarm.

Keywords: swarm control; distributed swarm; dynamic task planning; task assignment; event-trigger

Citation: Qin, B.; Zhang, D.; Tang, S.; Wang, M. Distributed Grouping Cooperative Dynamic Task Assignment Method of UAV Swarm. *Appl. Sci.* **2022**, *12*, 2865. https://doi.org/10.3390/app12062865

Academic Editors: Vincent A. Cicirello and Seong-Ik Han

Received: 18 January 2022
Accepted: 9 March 2022
Published: 10 March 2022

Publisher's Note: MDPI stays neutral with regard to jurisdictional claims in published maps and institutional affiliations.

1. Introduction

The advances of intelligent autonomous systems have led UAV swarm technology and its application to the current scientific research hotspot [1,2]. Cooperative task assignment stands out as an essential component and a precondition of task accomplishment and autonomous control of UAV swarm systems [3].

Cooperative task assignment is to assign a considerable number of different types of subtasks and their order to each UAV in the swarm while meeting the task requirements with UAV capabilities and the multiple constraints involved. In the past few decades, there have been several main sorts of assignment algorithms for UAV swarm cooperative task assignment: the heuristic algorithm and the market-based algorithm, for example [4]. Heuristic algorithms generally search a certain range of solution space in an acceptable time by simulating natural phenomena to obtain feasible solutions to optimization problems. Common methods include the genetic algorithm [5–8], the particle swarm optimization algorithm [9], the ant colony algorithm [10,11], and the wolf swarm algorithm [12]. For UAV swarms with distributed architecture, heuristic algorithms always need to obtain global information. This process consumes lots of communication and computing resources as well as time to achieve global consensus; it is not suitable to apply heuristic algorithms. Compared with heuristic algorithms, market-based approaches (such as contract network protocol), distinctively characterized by distributed computing of swarms, requires only local information of the swarm and have the advantages of flexibility, robustness, and high operation speed [13] and, hence, are more suitable for distributed UAV swarms. Meanwhile, with strong scalability, market-based approaches can well handle cooperative task assignments with complex constraints such as limited communication [14] and time windows [15] and have been applied to dynamic task assignments [16–18]. The algorithms adopted are a part of the assignment approaches for the task assignment in complex and changeable combat scenarios, which are usually modeled as constraints or objectives from different perspectives.

The crucial topics to be investigated in this field include cooperative task assignment with the trajectory coupled [19,20] as well as under dynamic resistant circumstances [21–23]. In [24], a scheme to assign tasks in UAV swarms based on the contract network protocol (CNP) is presented, in which an A* algorithm is applied in flight path planning and path length estimation with a no-fly zone and threat considered. Under this scheme, the coupling between task assignment and flight path planning can be solved. However, their work does not take into consideration either pop-up missions or UAV faults. A study [25] has proposed a novel model for UAV coalition and an algorithm derived from basic geometry that generates a path derived from the original Dubins curve for application in remote sensing missions of fixed-wing UAVs. Another study [26] proposed an unmanned air vehicle (UAV) swarm task and a resource dynamic assignment algorithm based on the task sequence mechanism. By establishing a task sequence, each UAV strictly separates the necessary task time and synchronization waiting time. For the newfound targets, each UAV quickly determines its available time period. According to the available time and task resources, an auction algorithm and a consensus algorithm are used to decompose the task assignment into the initial distributed assignment phase and the swarm consensus phase to develop real-time conflict-free task solutions for UAV swarms. However, their work does not take into account the communication topology and time constraints. In [27], a CNP-based approach to a multi-UAV task assignment is proposed, in which a flight path planning method based on PH curves is combined with cooperative particle swarm optimization (PSO), cooperative variables, and cooperative functions to achieve attack synchronization on certain targets. Nevertheless, it does not take into account no-fly zones, threats, communication topology, and time constraints in addition to the task reassignment in the case of UAV faults. On the basis of this, [27,28] introduced local communication constraints with communication distance and information hop times to determine whether other UAVs participate in the local task assignment; however, it neglected the communication constraint caused by the swarm specific topology, which is crucial for certain command structures so as to improve operational effectiveness.

Compared with [21–28], the problem investigated in this paper is the dynamic task assignment for the heterogeneous UAV swarm consisting of distributed subsets with specific topology. In this paper, common constraints such as time windows, the UAV capability model, as well as new constraints such as topology constraints are combined to build the complex model of dynamic task assignment. The key contribution of this paper is that it proposes a solution to rectify the problem of heterogeneous UAV swarm cooperative dynamic task assignments with specific hierarchical communication topologies and other multiple constraints. An extended-CNP-based distributed assignment approach is proposed, along with distributed heterogeneous UAV swarm executing reconnaissance and attack tasks as the main scenario. The swarm consists of several subsets. On the swarm discovering new targets, it firstly assigns each target to subsets according to communication topology, and then each subset assigns subtasks to the UAVs within the group. The modified artificial potential field method is adopted to preplan the threat avoidance flight path, and battlefield survivability and fuel consumption are introduced to describe the flight path's impact on the task assignment. Correspondingly, the consumption penalty and threat penalty functions can be designed so as to solve the coupling between task assignment and path planning. Meanwhile, constraints on time and cooperation are introduced to adjust the task executing sequence.

The rest of paper is organized as follows. The description of distributed grouping UAV swarm task assignment problem with multiple constraints is presented in Section 2. In Section 3, combined with hierarchical communication topologies of the UAV swarm, the distributed assignment algorithm based on extended contract network protocol is thoroughly addressed. In Section 4, the dynamic cooperative task assignment scheme based on an event-trigger strategy is proposed. Section 5 demonstrates the approach's effectiveness, both in uncertain environments and with UAV failure, by numerical examples, and the whole work is concluded in Section 6.

2. Problem Description

2.1. Mission Scenario Analysis

There are heterogeneous UAV swarms consisting of distributed subsets in the mission area. Each subset is composed of several reconnaissance UAVs, attack UAVs, and reconnaissance-attack UAVs. The procedure of task assignment includes target allocation and subtask assignment. The swarm accomplishes the initial assignment, and then the UAVs cooperatively perform tasks. If a UAV discovers a new target, it relays the target information to others within a limited range according to hierarchical communication topologies, which triggers dynamic task assignment and then updates each UAV's task sequence.

2.2. Hierarchical Communication Topology

The UAV swarm is divided into several distributed subsets. Communication exists within each subset and among subsets, hence hierarchical communication topology is established. In the application process, the swarm can cooperate to assign and complete tasks through the two-layer mechanism of inter-group cooperation and intra-group coordination based on hierarchical communication topology. The structure of the distributed subsets adopted integrates the advantages of both centralized structure and fully distributed structure to realize "global centralization and local autonomy", which avoids the problems of low redundancy and the heavy central load of the fully centralized structure as well as the disadvantages of high individual capability requirements, communication complexity, and the command conflicts of the fully distributed structure. The structure conforms to the actual combat scenario as well as the development status of UAV swarm technology and will become more normalized and practical [29–32].

The algebraic graph theory is used to describe the internal interaction of the UAV swarm system. Assume that the UAV swarm has N UAVs, and each UAV is regarded as a node, then the communication relationship is seen as an edge. A directed graph $G = \{V, E, W\}$ which consists of the node set $V = \{v_1, v_2, \ldots, v_N\}$, the edge set $E \subseteq \{(v_i, v_j) : v_i, v_j \in V, i \neq j\}$ and the adjacency matrix $W = [w^{ij}] \in \mathbb{R}^{N \times N}$, with non-negative entries w^{ij}. The entries in W are defined with $w^{ij} = 1$ for $(v_i, v_j) \in E$ and $w^{ij} = 0$ otherwise. In addition, $w^{ii} = 0$ for all $i = 1, 2, \cdots, N$. The neighbor set of node v_i is described as $N_i = \{v_j : (v_i, v_j) \in E\}$.

There is a top leader, N_m group leaders, and N_f followers in UAV swarms with distributed subsets, as shown in Figure 1. Each subset i has N_{fi} followers. The top leader is the highest leader node and leads the initial task assignment. The group leader, which obtains information of each UAV in the subset, is the leader node of the subset, participates in target allocation on behalf of the subset, and leads subtask assignment; the followers are members of the subset and participate in subtask assignment and execution.

Each node is numbered in the UAV swarm, in which the top leader index is 1, the indexes of group leaders are $i = 2, \ldots, N_m + 1$, and the indices of followers are $i = N_m + 2, \ldots, N_m + N_f + 1$. The adjacency matrix $W \in \mathbb{R}^{N \times N}$ has the following form:

$$W = \begin{bmatrix} 0 & W^{TM} & 0 \\ W^{MT} & W^{M0} & W^{MF} \\ 0 & W^{FM} & W^{F0} \end{bmatrix}_{N \times N} \tag{1}$$

where $W^{MT} \in \mathbb{R}^{N_m \times 1}$ and $W^{TM} \in \mathbb{R}^{1 \times N_m}$ indicate the communication topology among the top leader and group leaders, $W^{M0} \in \mathbb{R}^{N_m \times N_m}$ expresses the communication topology among group leaders, the communication topology among group leaders, and their followers are denoted as $\boldsymbol{W}^{M0} \in \mathbb{R}^{N_m \times N_m}$, while $\boldsymbol{W}^{F0} \in \mathbb{R}^{N_f \times N_f}$ is denoted as the communication topology among followers. Assuming that no direct communication exists between each follower and the top leader, the followers of each subset only communicate within UAVs in the subset:

$$
\begin{aligned}
\boldsymbol{W}^{FM} &= diag\left\{\boldsymbol{W}^{FM_1}, \boldsymbol{W}^{FM_2}, \ldots, \boldsymbol{W}^{FM_{N_m}}\right\} \\
\boldsymbol{W}^{MF} &= diag\left\{\boldsymbol{W}^{MF_1}, \boldsymbol{W}^{MF_2}, \ldots, \boldsymbol{W}^{MF_{N_m}}\right\} \\
\boldsymbol{W}^{F0} &= diag\left\{\boldsymbol{W}^{F_1}, \boldsymbol{W}^{F_2}, \ldots, \boldsymbol{W}^{F_{N_m}}\right\}
\end{aligned}
\tag{2}
$$

where $\boldsymbol{W}^{FM_i} \in \mathbb{R}^{N_{fi} \times 1}$ and $\boldsymbol{W}^{MF_i} \in \mathbb{R}^{1 \times N_{fi}}$ express the communication topology among group leader and its followers in subset i, and $\boldsymbol{W}^{F_i} \in \mathbb{R}^{N_{fi} \times N_{fi}}$ is the topology among followers in subset i.

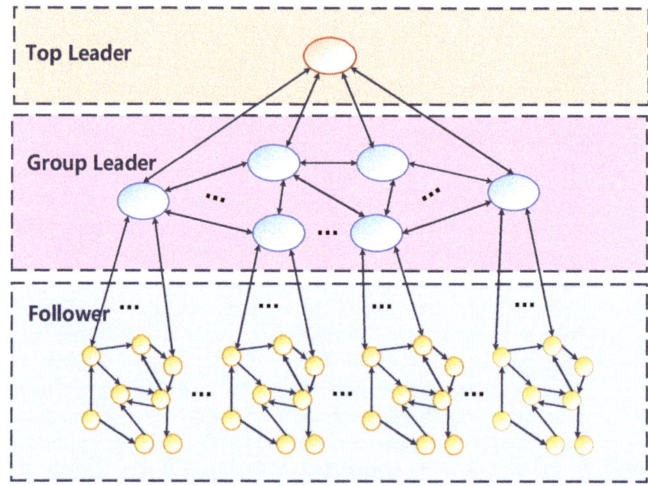

Figure 1. The hierarchical communication topology of UAV swarms with a distributed group.

2.3. UAV and Payload Model

Denote UAV set $V = \{V_i \mid i = 1, 2, \ldots, N\}$, in which any UAV V_i can be described as a seven-element combination $< X_i, h_i, C_i, P_i, F_i, D_{i,max}, N_{vi,max} >$. $X_i = (x_i, y_i, z_i, v_i, \varphi_i, \psi_i)$ represents the kinetic parameters of UAV V_i, including position, velocity, flight path angle, and flight heading angle; $h_i \in [0, 1]$ represents the survivability of UAV V_i, and $h_i = 0$ indicates V_i is destroyed or has encountered faults and then completely loses its mission capability; $C_i = \{C_{i1}, C_{i2}, \ldots, C_{inCi}\}$ expresses indices of UAVs that communicate with V_i; $P_i = \{P_{i,scout}, P_{i,attack}\}$ is the executable task type of V_i, where $P_{i,scout} \in \{0, 1\}$, and $P_{i,attack} \in \{0, 1\}$ means reconnaissance capability and attack capability, respectively. When the capability is available, the value is 1, otherwise 0; F_i is the fuel consumption rate per unit air-range of V_i; $D_{i,\max}$ is the maximum air-range; $N_{vi,\max}$ is the maximum number of executable tasks of V_i.

The condition that a reconnaissance UAV discovers and confirms the threat or target is that it is located in the detection area of the reconnaissance payload, as shown in Figure 2a. Reference [33] gives the typical mathematic model of the detection area. The condition that an attack UAV can strike a target is that the target is located in the available area of the attack payload, as shown in Figure 2b. References [33,34] give the typical mathematic model of the available area.

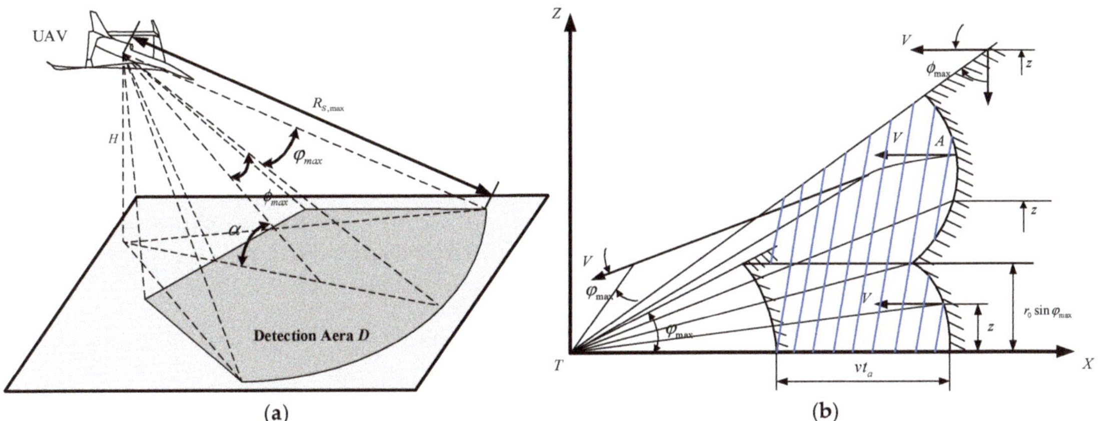

Figure 2. (**a**) The detection area of reconnaissance payload; (**b**) the available area of attack payload.

2.4. Threat Model

Any threat can be described by a five-element combination $< q_{o,i}, R_{o,i}, R_{a,i}, p_{o,i}, F_{o,i} >$, where $q_{o,i}$ is the position vector of the threat O_i center, $R_{o,i}$ is the no-fly zone radius of threat O_i, and $R_{a,i}$ is the impact radius of threat O_i. When a UAV flies into the no-fly zone, it will be destroyed; when the UAV enters the impact area, it will be affected by the threat. $p_{o,i} \in [0,1]$ expresses the estimation of threat impact. $F_{o,i} \in \{0,1\}$ indicates the detected status flag of the target, where the value 0 means the target is undetected; otherwise, the value is 1.

The estimation $p_{o,i}(q)$ of the threat O_i impact on any point in space is denoted as

$$p_{o,i}(q) = \begin{cases} 1, \ 0 < \|q - q_{o,i}\| < R_{o,i} \\ A_i(q, q_{o,i}), \ R_{o,i} \leq \|q - q_{o,i}\| \leq R_{a,i} \\ 0, \ \|q - q_{o,i}\| > R_{a,i} \end{cases} \tag{3}$$

where $A_i(q, q_{o,i}) \in (0,1)$, which is a function of the distance between the point and the threat center, represents the effect evaluation within the threat range; for the convenience of the study, $A_i(q, q_{o,i})$ can be taken as a constant value. When the UAV approaches threat O_i and enters the impact range, its survivability will decrease. We assume the survivability of UAV V_j is h_j and the survivability becomes h'_j under the impact of threat O_i, of which the process can be expressed as

$$h'_j = h_j \cdot (1 - A_i) \tag{4}$$

2.5. Dynamic Task Allocation Problem

We adopt a seven-element combination $\{V, Sg, T, O, M_t, R, C\}$ to describe the dynamic task assignment problem. $V = \{V_1, V_2, \ldots, V_N\}$ is the set of UAVs and N represents the number of UAVs in the swarm; $S_g = \{S_{g1}, S_{g2}, \ldots, S_{gNG}\}$ is the set of subsets and N_G represents the number of UAV subsets in the swarm; $T = \{T_1, T_2, \ldots, T_{NT}\}$ is the set of targets, where N_T is the number of targets; $O = \{O_1, O_2, \ldots, O_{Nobs}\}$ is the set of obstacles, where N_{obs} is the number of obstacles; $M_t = \{M_{t1}, M_{t2}, \ldots, M_{t,Ntype}\}$ is the task type set of each target, where N_{type} is the number of the types. For the "Reconnaissance-Attack" mission scenario, the task type set includes the reconnaissance and attack of two elements, which can be expressed as $M_t = \{Scout, Attack\}$; $R = \{R_1, R_2, \ldots, R_{Ntype}\}$ is the set of the maximum task values for the "Reconnaissance-Attack" mission scenario $R = \{R_S, R_A\}$; C is the set of multiple constraints, mainly including UAV capacity constraint, time window constraint, sequential constraint, and cooperation constraint. These constraints are described as follows:

Definition 1 (UAV capability constraints). *The UAV capability constraints are mainly reflected in three aspects: the maximum range, the executable task types, and the maximum number of executable tasks.*

(a) (The maximum range constraint) Assume that the initial state of UAV V_i is s_{i0}, and the task sequence is $Seq_i = \{s_{i1}, s_{i2}, \ldots, s_{i,Ns,i}\}$, where $N_{s,i}$ is the number of tasks to be executed. From Section 2.3, the maximum range of UAV V_i is $D_{i,max}$ and the maximum range constraint can be expressed as

$$\sum_{j=1}^{N_{s,i}} L(s_{ij}, s_{i,j-1}) \leq D_{i,max} \tag{5}$$

where $L(s_{ij}, s_{i,j-1})$, which relates to task $s_{i,j-1}$ and task s_{ij}, represents the air-range from the position of task $s_{i,j-1}$ to the position of task s_{ij}. That means the whole air-range couples with the task sequence.

(b) (The executable task type constraint) In the process of the swarm cooperative task execution, different sorts of UAVs perform different types of tasks. There is a mapping between the UAV capability $P_i = \{P_{i,scout}, P_{i,attack}\}$ and $M_t = \{Scout, Attack\}$, which can be expressed as

$$P_i = \{P_{i,scout}, P_{i,attack}\} = \begin{cases} \{1, *\} \rightarrow Scout \\ \{*, 1\} \rightarrow Attack \end{cases} \tag{6}$$

(c) (The task number constraint) The payload number and energy carried by UAVs have limits; thus, it is necessary to restrict the maximum number of tasks performed by the UAV. Assuming that the number of tasks assigned to UAV V_i is $N_{vt,i}$ and the upper limit is $N_{vi,max}$, the constraint is expressed as

$$N_{vt,i} \leq N_{vi,max} \tag{7}$$

Definition 2 (Sequence constraint). *If there is a specific execution order between subtasks $Task_i$ and $Task_j$, there is sequence constraint between $Task_i$ and $Task_j$. Reference [9] gives the concrete model of sequence constraint.*

Definition 3 (Time window constraint). *The start time when task $s_{i,j}$ in $Seq_i = \{s_{i1}, s_{i2}, \ldots, s_{i,Ns,i}\}$ is performed by UAV V_i needs to be guaranteed to be in the time window $t_{si,j} \in [t_{b,j}, t_{e,j}]$, where $t_{b,j}$ is the earliest start time and $t_{e,j}$ is the latest one. The start time $t_{si,j}$ has relations with the last task and the preplanned flight path of the UAV. Suppose $t_{e,j-1}$ the time when UAV accomplishes the last task and the preplanning air-range from $s_{i,j-1}$ to $s_{i,j}$, then the time window constraint can be expressed as*

$$\begin{cases} t_{si,j} \leq t_{e,i} \\ t_{si,j} = \max\left\{t_{b,i}, t_{e,j-1} + \frac{L_i}{V}\right\} \end{cases} \tag{8}$$

Definition 4 (Cooperation constraint). *For the process of several UAVs cooperatively performing reconnaissance or attack task $Task_j$, the expected number of participants $N_{pe,j}$ is introduced, which means that $Task_j$ can be accomplished by $N_{pe,j}$ UAVs at most. The actual number of participants is $N_{p,j}$ and has*

$$0 \leq N_{p,j} \leq N_{pe,j} \tag{9}$$

Definition 5 (The dynamic task allocation problem). *The objective is to find the best assignment. To be more concise is to optimize the swarm's reward B_s during the task execution, such that:*

$$
\begin{aligned}
\hat{B}_s &= \max B_s(V, S_g, T, O, M_t, R) \\
&= \max \sum_{i=1}^{N} \sum_{j=1}^{N_{s,i}} B_{ij}(h_{i,est}, L_{ij}, t_{task_j})
\end{aligned}
$$

subject to the constraints:

$$
\left\{
\begin{array}{l}
\sum\limits_{j=1}^{N_{s,i}} L(s_{ij}, s_{i,j-1}) \leq D_{i,max} \\[2mm]
P_i = \{P_{i,scout}, P_{i,attack}\} = \left\{ \begin{array}{l} \{1, *\} \rightarrow Scout \\ \{*, 1\} \rightarrow Attack \end{array} \right. \\[2mm]
N_{vt,i} \leq N_{vi,max} \\[2mm]
t_{si,j} \leq t_{e,i} \\[2mm]
t_{si,j} = \max\left\{ t_{b,i}, t_{e,j-1} + \frac{L_j}{V} \right\} \\[2mm]
0 \leq N_{p,j} \leq N_{pe,j}
\end{array}
\right. \quad , \forall i \in V
$$

where the value of B_{ij} indicates the reward of UAV i performing the task j; the survivability estimation is $h_{i,est}$, which means more loss to the UAV performing the task; the air-range performing $Task_j$ is L_{ij}; the start time of task j is t_{task_j}.

Due to the objective function coupling with the process of contract net protocol, the detailed expression is designed in Section 3.2, where the bidding function of the market mechanism is introduced.

3. Task Assignment Algorithm Based on Extended CNP

This section designs a distributed dynamic task assignment algorithm based on extended contract network protocol. The core of contract net protocol (CNP) is to simulate the "bid–win" market mechanism and realize the optimization of task assignment based on the interaction of individuals. The classical CNP is not suitable for sequential task assignments with multiple constraints under multiple rounds, which means that it needs extending according to complex task constraints.

3.1. Distributed Multi-Constraint Dynamic Task Assignment Algorithm

The algorithm includes four steps: target information release, bidding scheme generation, bid winning authorization, and task execution. The process is shown in Figure 3. The following describes the algorithm process in turn.

Step 1: Release the information of targets

Assume that the UAV detects a sudden target T_k and reports the target information to the subset leader UAV V_i, which becomes the tendering UAV on behalf of the subset, and sends a bidding invitation to each subset in the local communication network.

According to the hierarchical communication topology, the set of other subset leaders interacting with V_i is $V_{p,i} = \{V_j | w_{ij} = w_{ji} = 1, j \in [2, 1 + N_m] \cup j \neq i\}$. $V_{p,i}$ composes the potential bidders of assignment for target T_k. Tendering UAV V_i releases tendering information I_i to each UAV in set $V_{p,i}$, and I_i is expressed as

$$
I_i = (V_i, T_k, \{Task_k\}, t_{now}, H_b) \tag{10}
$$

where V_i is the index of tendering UAV; T_k is the index of the sudden target; $\{Task_k\}$ is denoted as the subtask set of target T_k; t_{now} is the releasing time; H_b is the information state flag, of which value 0 means no relay, otherwise 1. The information structure of $Task_k$ can be denoted as:

$$Task_k = \left\{ M_{t,k}, q_{t,k}, TimeBar_k, R_{x,k}, N_{pe,j}, Th_k \right\} \tag{11}$$

where $M_{t,k}$ means task type; $q_{t,k}$ is the position coordinates; $TimeBar_k = [t_{b,k}, t_{e,k}]$ is the time window; $R_{x,k}$ is the maximum task value; $N_{pe,j}$ is the expected number of participants; Th_k is the negotiation threshold.

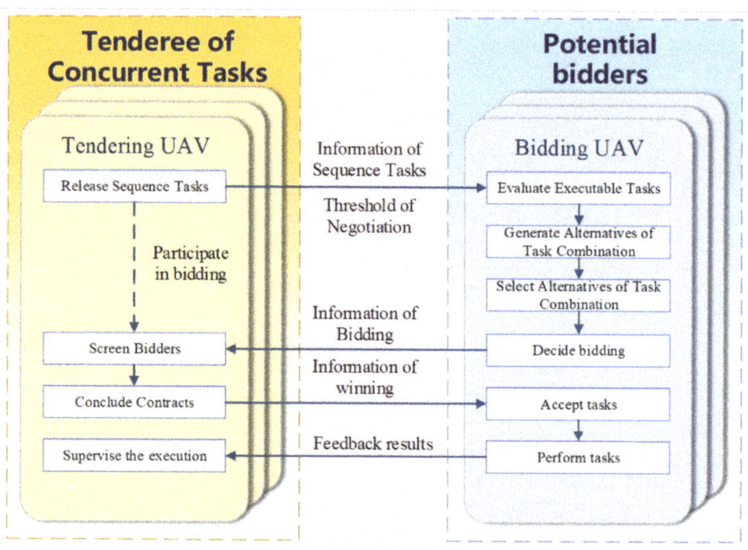

Figure 3. The process of task assignments based on extended CNP.

As attached information during the subtask $Task_k$ assignment process, negotiation threshold Th_k is applied to preselect bidders from the potential bidder set $\{V_{p,i}\}$, reducing the negotiation scale as well as the consumption of communication resources and improving assignment efficiency. To adjust the threshold adaptively, the reward of subtask $Task_k$ in performing the subset to which UAV V_i belongs is selected as Th_k. If the bidder's bidding value is greater than Th_k, it indicates that the swarm efficiency will be optimized and improved after $Task_k$ is assigned to the bidder.

In the process of information release, different UAV groups with intersections of local communication topologies may discover the same target, causing multiple bidding UAVs to release information at the same time in their respective local communication topologies, resulting in system conflict and resource waste. In order to avoid this problem, it is necessary to ensure that different UAV subsets reach a consensus on tendering UAVs and tendering information, which is realized by Algorithm 1.

In Algorithm 1, "StopCmd" means to abort the negotiation command, and the operation "." represents the reference to elements in the information structure. For any UAV V_i in the local communication topology, the tendering information related to subtask $Task_k$ will be released to its directly connected UAV if it is not empty and has not been transmitted (Line 1–4). Meanwhile, the UAV can also receive the tendering information related to $Task_k$ transmitted by others directly connected with it (Line 6–11).

Algorithm 1: Release tendering information and achieve information consensus.

1:   if I_i is not empty
2:   if $I_i.H_b = 0$
3:   broadcast I_i to $\left\{ V_{p,i} \right\}$
4:   endif
5:   endif
6:   for $j = 1$ to number of tendering $Task_k$ invitation received
7:   if $I_i.Task_k.Th_k < I_j.Task_k.Th_k$
8: $I_i = I_j$
9:   broadcast StopCmd to $\left\{ V_{p,i} \right\} - \left\{ V_{p,j} \right\}$
10:   endif
11: endfor

If UAV V_i finds that the negotiation threshold of UAV V_j about $Task_k$ is greater, V_i saves the tendering information I_j and issues commands to other UAV groups interacted with V_i but not interacted with V_j to stop the negotiation so as to make them exit the assignment dominated by UAV V_j.

After Step 1, the tendering information released in the local communication network achieves consensus, that is $I_i = I_j$.

Step 2: Generate bidding scheme

On the basis of the target information and various constraints, each potential bidder first determines the bidding scheme for the target of the group, including whether the subset participates in the bidding for target T_k, subtasks that can be performed and the corresponding UAVs, and the reward for performing each subtask, and so forth.

Through the contract network in subsets, each subtask of target T_k is preassigned to form the bidding scheme for target T_k. The specific process is as follows:

① A subset leader UAV V_j releases subtask information to each follower UAV. In order to improve the assignment efficiency and shorten the time, the task concurrency mechanism is introduced to publish the information of each subtask at the same time;

② For subtask $Task_k$, the follower UAV V_m judges whether it has the corresponding type of payload, whether the number of payloads can meet the task execution requirements, and whether it meets the time window $TimeBar_k$ according to the subtask information. If any constraint is not satisfied, UAV V_m will not bid for subtask $Task_k$;

③ If the above constraints are met, UAV V_m preplans the flight path of $Task_k$ based on the modified artificial potential field (MAP) method and substitutes the preplanned range L_{mk} and survivability estimation $h_{m,est}$ into the individual bidding function to calculate the bidding value B_{mk};

④ Compare the bidding value B_{mk} with the subtask negotiation threshold Th_k. If $B_{mk} < Th_k$, it means that the system efficiency has not been improved when UAV V_m is used to perform subtask $Task_k$, and V_m actively abandons pre-assignment bidding for $Task_k$; otherwise, V_m participates in the bidding;

⑤ Each follower UAV respectively performs steps ②–④ to complete the pre-assignment of each subtask. The subset leader UAV V_j generates the subset's bidding scheme for target T_k according to the pre-assignment results.

The above is the basic process of bidding scheme generation. Due to the existence of multi-UAV cooperatively performing subtasks and the introduction of the task concurrency mechanism, the generation of the bidding scheme needs to solve a "multi-participants–multi-tasks" assignment optimization subproblem.

In order to solve the subproblem, an assignment mechanism based on the contract network within subsets is constructed. For the subtasks that need to be executed by multiple UAVs, a cooperator determination mechanism is introduced to complete the task allocation; for the sequential subtasks with time windows, a sequential task selection mechanism is introduced. Details are as follows:

(1) The determination mechanism on cooperators

The process of the determination mechanism on cooperators is shown in Figure 4. To simplify the assignment process, give the tenderee the initiative, expand the set of bidders, and provide more feasible pre-assignment schemes, the tenderee bidding mechanism is introduced, that is, when the subtask tenderee (subset leader UAV V_j) meets conditions such as capability constraints and time constraints, it can participate in bidding.

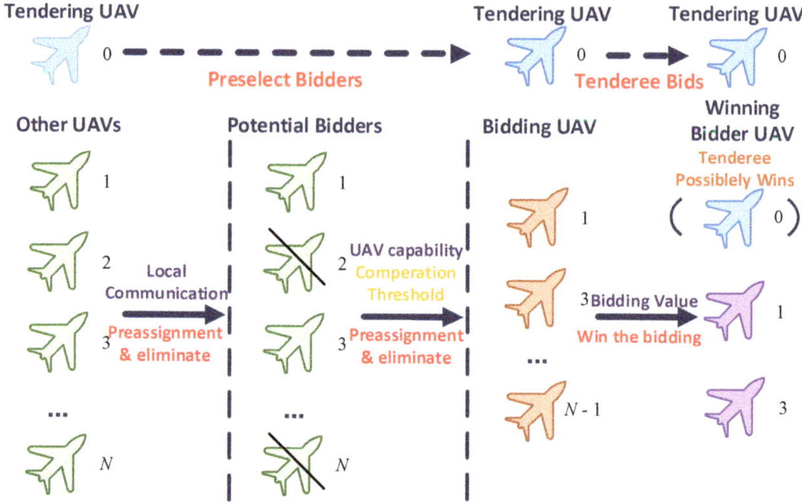

Figure 4. The determination mechanism on cooperators.

If V_j meets the constraints for performing $Task_k$ and decides to bid with the bidding value B_{jk}, then Th is chosen as B_{jk}. UAV V_j sends $Task_k$ information and negotiation threshold B_{jk} to each follower UAV. The bidding value of follower UAV V_m for $Task_k$ is B_{mk}. If $B_{mk} > B_{jk}$, V_m decides to bid and feedback B_{mk} to V_j. At the time, the set of all bidders participating in the $Task_k$ assignment is

$$V_{b,k} = \left\{ V_m \middle| w_{jm} = w_{mj} = 1, j \neq m \,; B_{mk} > B_{jk} \right\} \cup \left\{ V_j \right\}$$

If tenderee V_j does not participate in the bidding, i.e., $B_{jk} = \varnothing$,

$$V_{b,k} = \left\{ V_m \middle| w_{jm} = w_{mj} = 1, j \neq m \,; B_{mk} > 0 \right\}$$

To sum up, after preselection of negotiation threshold, all bidders participating in the assignment for $Task_k$ can be represented as

$$V_{b,k} = \begin{cases} \left\{ V_m \middle| w_{jm} = w_{mj} = 1, j \neq m; B_{mk} > 0 \right\}, B_{jk} = \varnothing \\ \left\{ V_m \middle| w_{jm} = w_{mj} = 1, j \neq m; B_{mk} > B_{jk} \right\} \cup \left\{ V_j \right\}, B_{jk} \neq \varnothing \end{cases} \tag{12}$$

Consider the cooperative constraints of subtask $Task_k$, assume $Task_k$ requires the participation of $N_{p,k}$ UAVs, and quickly sort the bidding value set $\left\{ B_{jk} \right\}$. The greedy algorithm is used to select the $N_{p,k}$ largest bidding values in the set to form the winning bidding value set (reward set) B_k

$$B_k = \left\{ B_{(1),k}, \, B_{(2),k}, \, \dots, \, B_{(Np,k),k} \right\} \tag{13}$$

where $B_{(x),k}$ represents the *x-th* order element of $\{B_{mk}\}$ in descending order. The winner set $Winner_k$ is correspondingly:

$$Winner_k = Index(B_{(1),k}, B_{(2),k}, \ldots, B_{(Np,k),k}) \tag{14}$$

(2) Sequential task selection mechanism

Take the sequential task assignment process of a single follower UAV as an example. After receiving the task information, the bidder determines the executable tasks according to the negotiation threshold, UAV capability constraints, and the execution sequence and combines them to generate an alternative sequence without a time window conflict. Based on the bid winning situation, the optimal sequence is selected as the final signing one. The operation process is shown in Figure 5.

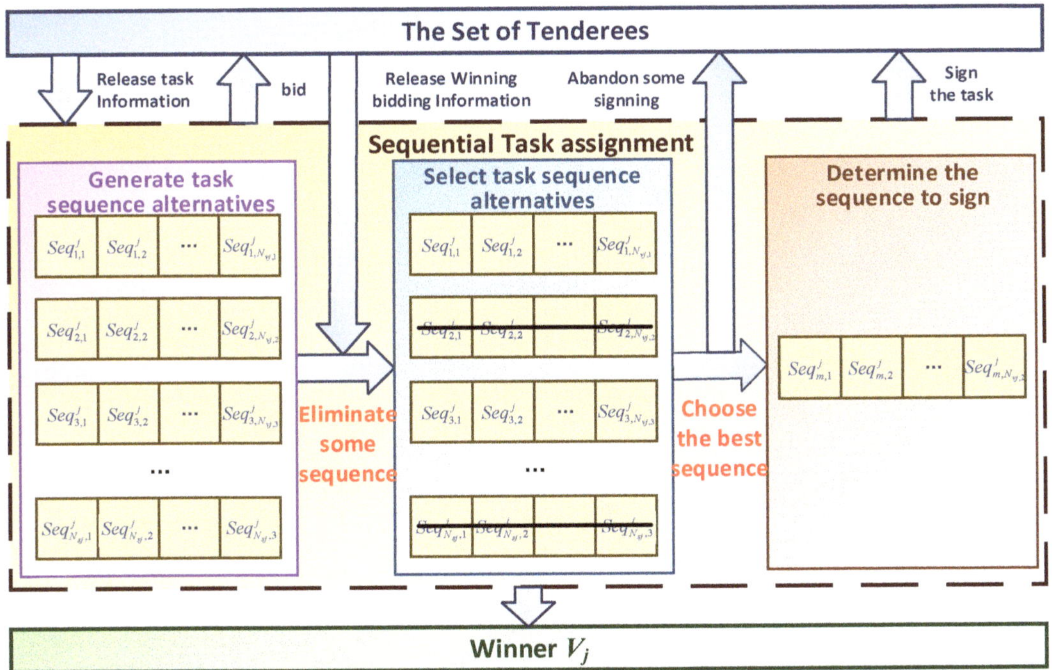

Figure 5. The process of task assignment based on extended CNP.

Assume that the concurrent task set $Task^{new} = \{Task_1^{new}, \ldots, Task_{Ntask}^{new}\}$ is received by the follower UAV V_m. The set of tasks that V_m can execute and determine to participate in bidding is expressed as

$$Task^{a,m} = \{Task_k^{a,m} \mid k = 1, 2, \ldots, N_{a,m}\} \subseteq Task^{new} \tag{15}$$

where $N_{a,m}$ is the element number of $Task^{a,m}$. Denote $Comb_n(N_{a,m}, N_{vm,n})$ as the combination which consists of random $N_{vm,n}$ elements from set $\{1, 2, \ldots, N_{a,m}\}$, where $N_{vm,n} \leq N_{a,m}$.

Definition 6 (Task Sequence Alternative). *If $\forall k \in Comb_n(N_{a,m}, N_{vm,n})$, $\forall \bar{k} \in Comb_n(N_{a,m}, N_{vm,n}) - \{k\}$, there is $TimeBar_k \cap TimeBar_{\bar{k}} = \varnothing$. Then, $Comb_n(N_{a,m}, N_{vm,n})$ is called a sequence alternative. $TimeBar_k$ is the time window of subtask $Task_k^{a,m}$. The corresponding task sequence alternative is defined as*

$$Seq_n^m = \{Task_k^{a,m} \mid k \in Comb_n(N_{a,m}, N_{vm,n})\} \;, n \leq N_{sm}, Seq_n^m \subseteq Task^{a,m} \tag{16}$$

where N_{sj} is the number of task sequence alternatives.

After the bidding is completed, the set of losing bidding tasks is denoted by $Task^{de,m}$. After the sequences containing losing bidding tasks are eliminated, the set composed of the remaining sequence alternatives is

$$SeqSet_m = \left\{ Seq_n^m \middle| n \leq N_{sm}; \forall Task_k^{a,m} \in Seq_n^m, Task_k^{a,m} \notin Task^{de,m} \right\} \tag{17}$$

Assuming that the bidding value of V_m for any subtask $Task_k^{a,m}$ in Seq_n^m is B_{mk}, where $Seq_n^m \in SeqSet_m$, the sum of bidding value for each task in Seq_n^m is

$$B_{m,n} = \sum_k B_{m,k}, \ k \in Comb_n(N_{a,m}, N_{vm,n}) \tag{18}$$

and the final signing sequence of follower UAV V_m is

$$Seq^{win,m} = Seq_{n_{best}}^m, n_{best} = \underset{n}{\arg\max}(B_{m,n}) \tag{19}$$

Through the above mechanism, the subtask pre-assignment scheme of the subset in which V_j is located is generated, that is, the bidding scheme of the subset

$$BidSch_j = \left\{ Seq^{win,m} \middle| m \in G_j \right\} \tag{20}$$

After that, UAV V_j calculates the bidding function of the subset according to the bidding scheme and bids to the tenderee V_i.

Step 3: Determine the winners

After the bidding scheme generation and bidding application in Step 2, the tendering UAV V_i selects the scheme of which the efficiency is greatest according to the bidding value of each group so as to determine which subset executes subtasks of target T_k.

The set of bidding values of subsets is $\left\{ B_{g,jk} \right\}$. Apply the greedy algorithm and select the greatest value and corresponding subset as the winning value (reward) $B_{g,k}$ and the winner $Winner_{g,k}$. The process can be expressed mathematically as

$$B_{g,k} = \max\{B_{g,jk}, Winner_{g,k} = \underset{j}{\arg\max}(B_{g,jk}) \tag{21}$$

After determining the winner, UAV V_i authorizes the subset $Winner_{g,k}$ to perform each subtask of target T_k.

Step 4: Perform the tasks

The winner subset formally authorizes each follower UAV in the subset to perform the subtasks, and the pre-assignment scheme in Step 2 becomes the formal assignment scheme. Follower UAV V_m will add the signing subtask sequence $Seq^{win,m}$ to its task sequence to be executed, that is:

$$A_m = A_m \oplus Seq^{win,m} \tag{22}$$

where A_m is the task sequence to be executed, $\oplus$ expresses the operation that it adds the sequence $Seq^{win,m}$ to the sequence A_m. Each UAV performs each task according to sequence A_m and the preplanning flight path in the specific time window.

The pseudo-code of the whole algorithm 2 process described above is as follows:

Algorithm 2: The extended Contract Network Protocol.

Input: Task assignment combination $\{V, Sg, T, O, M_t, R, C\}$; Topology G_0; time t_{now}
Output: Task execution sequence set $\{A_m\}$

1:   /* **Step 1: Release the information of targets** */
2:   for i from the first index of leader UAVs to the last one
3:      if V_i is a target tenderee UAV
4:         $\{V_{p,i}\}$ = detTargetPotentialBidders(G_0,C)   /* determine the set of potential bidder subsets. */
5:         for k = 1 to number of targets
6:            if T_k is tendered by leader UAV i
7:              $\{Task_k\}$ = generateSubtasks(T_k)   /* Tenderee UAV i generates subtask information. */
8:            end if
9:         end for
10:         I_i = releaseTargetInfo $(\{T_k\}, \{Task_k\}, t_{now}, \{V_{p,i}\})$ /* Tenderee UAV i releases target information. */
11:      end if
12:   end for
13:   /* **Step 2: Generate bidding scheme** */
14:   for j from the first index of leader UAVs to the last one
15:      for i from the first index of tenderee leader UAVs to the last one
16:         if V_j belongs to $\{V_{p,i}\}$
17:            Receive I_i and broadcast it to V_j 's followers
18:         end if
19:      end for
20:      for m from the first index of followers of V_j to the last one
21:         $\{Task^{a,m}\}$ = checkTaskConstaints($\{I_i.Task_k\}$)   /* Select the executable subtasks from $\{Task_k\}$ */
22:         $\{Seq_n^m\}$ = combineTasks($\{Task^{a,m}\}$)   /* Generate the set of alternative sequence */
23:         $\{B_{m,n}\}$ = biddingFuncCal($\{SeqSet_m\}$)   /* Calculate the bidding function of each sequence */
24:         $Seq^{win,m} = Seq_{n_{best}}^m$, $n_{best} = \underset{n}{\mathrm{argmax}}(B_{m,n})$ /* Determine the sequence to execute */
25:      end for
26:      $B_{g,jk} = \sum\limits_{m} \max(B_{m,n})$
27:   end for
28:   /* **Step 3: Determine the winners** */
29:   for i = 1 to number of tenderee leader UAV i
30:      for k = 1 to number of targets tendered by leader UAV i
31:         $B_{g,k} = \max\{B_{g,jk}\}$, $Winner_{g,k} = \underset{j}{\mathrm{argmax}}\left(B_{g,jk}\right)$ /* Determine winner of tendering for Target k */
32:      end for
33:   end for
34:   /* **Step 4: Perform the tasks** */
35:   for n = 1 to number of subsets
36:      for m = 1 to number of follower UAV m of Subset n
37:         $A_m = A_m \oplus Seq^{win,m}$ /* Add the signing subtask sequence to task execution sequence A_m. */
38:      end for
39:   end for

Analyze the worst time complexity of the algorithm and take the case of task assignment by the top leader as an example:

The time complexity of the top leader issuing N_T targets with subtasks bidding information to N_G subset leaders is $O(N_T N_G)$, the complexity of the initial evaluation of 4 types of constraints presented in Section 2.5 by N_G subset leaders is $4O(N_T N_G)$, the complexity of the bidding of potential bidder subsets $\{V_{p,i}\}$ is $O(K)$, where K is the number of elements in $\{V_{p,i}\}$; after K potential bidder subsets are determined, each potential bidder subset needs to pre-assign the subtasks and generate the bidding scheme of the subset.

Assume that each target has N_t subtasks, each subtask needs the most N_p participants and each subset consists of $(N_f + 1)$ UAVs. Each subset will generate the target scheme based on the CNP within the subset. The time complexity of the subset leaders issuing N_t subtasks to their N_f followers is $O(N_T N_t N_f)$. The sequential task selection is an arrangement–

combination problem, of which the time complexity is $N_T \bullet O(N_t{}^3)$, and the time complexity of bidding to the subset leader in each subset is $O(N_T N_t N_f)$. Subset-leader-selecting cooperators can be regarded as a sorting problem, and the quick sort algorithm can be adopted so that the time complexity is $N_T N_t \bullet O(N_p \log_2 N_p)$. Therefore, the time complexity of the subtask pre-assignment is $2O(N_T N_t N_f) + N_T \bullet O(N_t{}^3) + N_T N_t \bullet O(N_p \log_2 N_p)$.

After the bidding schemes of each subset are generated, the bidding winner subset will be determined. It is a quick sort process, so the time complexity is $N_T \bullet O(N_G \log_2 N_G)$.

To conclude, the whole worst time complexity is $2O(N_T N_t N_f) + N_T \bullet O(N_t{}^3) + N_T N_t \bullet O(N_p \log_2 N_p) + N_T \bullet O(N_G \log_2 N_G)$. It indicates that the algorithm is a polynomial one.

3.2. Bidding Function Design

As the core of extended CNP, the bidding function, which needs to be designed on the basis of the concrete task assignment problem in CNP, is a sort of objective function. According to the method process described above, the whole task assignment is divided into the target assignment and the subtask assignment. Both the target assignment and the subtask assignment are based on the extended CNP. Hence, the bidding function of each subset is designed for the target assignment, while the individual bidding function within the group is designed for the subtask assignment.

3.2.1. Individual Bidding Function

The individual bidding function B_{ij} consists of value function Re_{ij} and penalty function Pe_{ij}. Based on the analysis for the UAV swarm task assignment model in Section 2, the individual bidding function mainly relates to: ① the maximum task values, which is denoted in Section 2.5; ② the flight path; ③ the time windows and start time of each subtask.

Since the UAV task assignment couples with path planning, the bidding function needs to be designed in combination with the UAV flight path. By analyzing the scenario, it can be seen that the impact of the flight path on task assignment is mainly reflected in UAV survivability estimation $h_{i,est}$ and air-range L_{ij} (fuel consumption), as shown in Figure 6.

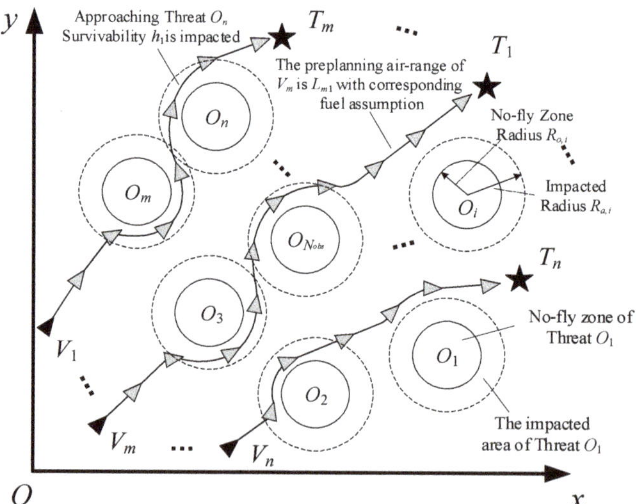

Figure 6. The coupling of the UAV mission assignment and path planning.

Supposing that UAV V_i participates in the bidding process of subtask $Task_j$ with time window $t_{task_j} \in [t_{s,j}, t_{e,j}]$, where t_{task_j} is the start time of task execution, of which the mathematical expression is given in Section 2.5, $t_{s,j}$ is the earliest start time, and $t_{e,j}$ is the latest one.

To evaluate the impact of threats $\{O_k \mid k = 1, 2, \ldots, N_{obs}\}$ on UAV V_i performing tasks, it introduces the survivability estimation of V_i, which can be expressed as

$$h_{i,est} = h_i \prod_{k=1}^{N_{app}} (1 - A_{k,app}) \tag{23}$$

where h_i is the current survivability of V_i, $A_{k,app}$ is the impact of threats $O_{k,app} \in \{O_k\}$ approached by V_i, $A_{k,app} \in (0,1)$, and N_{app} is the number of threats approached.

(1) Value function Re_{ij}

In the actual scenario, the value of UAVs performing any task is generally related to the start time when the UAV performs the task. The nearer the start time t_{task_j} to the time window $TimeBar_k$, the greater the value for UAV V_i to perform the task; otherwise, the smaller the value; hence, the item related to the start time t_{task_j} is introduced. The value function of V_i performing $Task_j$ is designed as

$$Re_{ij}(h_{i,est}, t_{task_j}) = h_{i,est} \cdot R_{x,j} \cdot e^{-\lambda \frac{t_{task_j} - t_{s,j}}{t_{e,j} - t_{task_j}}} \tag{24}$$

where $R_{x,j}$ is the maximum value of $Task_j$; λ is the attenuation factor of the start time t_{task_j} on the task value. Apparently, Re_{ij} is a monotonic increasing function of battlefield survivability $h_{i,est}$ and decreases monotonically with the start time t_{task_j}. When $h_{i,est} : 1 \to 0, t_{task_j} : t_{s,j} \to t_{e,j}$, there is $Re_{ij}: R_{x,j} \to 0$.

(2) Penalty function Pe_{ij}

The penalty function is composed of the consumption penalty and the threat penalty. Assuming that the per-unit fuel consumption of UAV V_i is F_i, the air-range performing $Task_j$ is L_{ij}, the estimation of survivability is $h_{i,est}$, and the penalty function is designed as

$$Pe_{ij}(h_{i,est}, L_{ij}) = \frac{1}{h_{i,est}} \cdot F_i L_{ij} \tag{25}$$

Obviously, Re_{ij} is a monotonic function of battlefield survivability $h_{i,est}$. When $h_{i,est} : 1 \to 0^+$, there is $Pe_{ij}: F_i L_{ij} \to +\infty$.

The penalty indicates the coupling impact of flight path planning on task assignment. On one hand, the fuel consumption of performing subtask $Task_j$ is directly related to the preplanned air-range L_{ij}; on the other hand, the more threats approached by UAV V_i during the task execution, the lower the survivability estimation $h_{i,est}$, which means more loss when the UAV performs the task.

Combining the value function and the penalty function, the individual bidding function B_{ij} of UAV V_i for $Task_j$ is denoted by

$$B_{ij}(h_{i,est}, L_{ij}, t_{task_j}) = Re_{ij} - Pe_{ij} = h_{i,est} \cdot R_{x,j} \cdot e^{-\lambda \frac{t_{task_j} - t_{s,j}}{t_{e,j} - t_{task_j}}} - \frac{1}{h_{i,est}} \cdot F_i L_{ij} \tag{26}$$

where a greater value of B_{ij} indicates the reward of performing the task. B_{ij} monotonically increases with $h_{i,est}$ and decreases with L_{ij} and t_{task_j}, which agrees with the actual scenario, which suggests that B_{ij} has practical significance.

3.2.2. Bidding Function of Each Subset

The bidding function of subsets is based on the individual function. For target T_j, supposing that the bidding value set is $\{B_1, B_2\}$ after the subtask assignment in subset G_i, B_1, B_2 respectively express the bidding value of subset G_i performing the reconnaissance subtask and the attack one. The bidding function of subsets is defined as the sum of elements in $\{B_1, B_2\}$, which is

$$B_{g,ij} = \sum B_1 + \sum B_2 \tag{27}$$

4. Dynamic Assignment Strategy Based on Event Triggering

4.1. Event Trigger Conditions for Dynamic Tasks

Due to the incomplete perception of the situation at the beginning and the occurrence of emergencies during the task execution, the initial assignment scheme needs to be adjusted. Therefore, the event trigger conditions are introduced to judge whether or not to carry out the dynamic task assignment. Event trigger conditions are the key to realizing dynamic task assignments, which need to be selected in combination with different scenarios. The dynamic task assignment process based on event triggering can be expressed as:

$$E : \text{if } g(s(t), s_{t,j}(t)) \geq 0, \text{then } Target_j \leftarrow TargetInfo_j^*, Task_{j,k} \leftarrow TaskInfo_{j,k}^*(k = 1,\ 2,\ \ldots,\ N_f)$$

where $g(s(t), s_{t,j}(t)) \geq 0$ is the triggering condition, $s(t)$ is the state of UAV (such as the position), $s_{t,j}(t)$ is the status of target T_j. $Target_j \leftarrow TargetInfo_j^*$ expresses the update of the set of targets, and $Task_{j,k} \leftarrow TaskInfo_{j,k}^*$ expresses the update of the set of subtasks.

(1) The unknown targets appear

The initial target assignment is based on the initial situation awareness information, which can only cover the known targets but is not guaranteed to cover all the targets in the mission area. After the reconnaissance UAV or reconnaissance-attack UAV V_i flies near the unknown target T_j during task execution, it will detect the target and obtains the target information $TargetInfo_{new,j}$ and then updates the target set and the subtask set. Event trigger conditions can be described as

$$g(s(t), s_{t,j}(t)) = DA(r_i, r_{nt,j}, R_{detect,i})$$

where $DA(r_i, r_{nt,j}, R_{detect,i})$ is defined as the reconnaissance payload constraint, which is the constraint on the relative position and angle relations between UAV V_i and target T_j. The details are in reference [34]. When target T_j is located in the detect area of the reconnaissance-attack UAV or reconnaissance UAV, $DA(r_i, r_{nt,j}, R_{detect,i}) \geq 0$; r_i is the position vector of UAV V_i; $r_{nt,j}$ is the position vector of target T_j; $R_{detect,i}$ is the detection range of UAV V_i. The event-triggering process can be expressed as

$$E_1 : \text{if } DA(r_i, r_{nt,j}, R_{detect,i}) \geq 0, \text{ then } Target_{N_{target}+j} \leftarrow TargetInfo_{new,j}, \quad \left(k = 1, 2, \ldots N_{newtask,j} \right)$$
$$Task_{N_{target}+j,k} \leftarrow TaskInfo_{New,j,k}$$

where $Target_{N_{target}+j} \leftarrow TargetInfo_{new,j}$ indicates the update of target set, N_{target} is the number of targets; $Task_{N_{target}+j,k} \leftarrow TaskInfo_{New,j,k}$ indicates the update of subtask set, $N_{newtask,j}$ is the number of subtasks of T_j.

(2) The UAV encounters faults and cannot perform tasks

During task execution, UAVs may encounter non-cooperative behaviors such as collision and attack and lose mission capability. Its tasks that have not been executed will be reassigned. Event triggering conditions can be described as

$$g(s(t), s_{t,j}(t)) = h_{i,b} - h_i$$

where h_i is the survivability of V_i; $h_{i,b}$ is the minimum survivability of V_i with normal capability and is selected according to the concrete scenario. The event-triggering process can be expressed as

$$E_2 : \text{if } h_{i,b} - h_i \geq 0, \text{ then } Task_{j,k} \leftarrow TaskInfo_{i,j,k} \ (k = 1,\ 2,\ \ldots,\ N_{i,j})$$

where $Task_{j,k} \leftarrow TaskInfo_{i,j,k}$ is the reassignment for the subtasks that V_i is to perform.

4.2. The Basic Process of Dynamic Task Assignment

Before describing the basic process of cooperative dynamic task assignment, the following assumptions are given for the scenario:

1. Based on the early situation awareness, several targets and pieces of threat information have been obtained, including target location, number of UAVs required to perform each subtask, threat location, and impact range;

2. Due to incomplete situational awareness, there are unknown targets and threats. After the UAV detects an unknown target or threat, it broadcasts the threat information, transmits the target information to the group leader, and triggers the assignment process.

Based on the scenario and assumptions above, the basic process of cooperative dynamic task assignment is shown in Figure 7. The concrete steps are as follows:

1. Assign the known targets. The top leader assumes the role of the tenderee, dominates the initial assignment, and releases the target information to each group leader based on the inter-group contract network.

2. Each group leader that has direct communication with the top leader receives the target information, releases the subtask information to their followers, and determines the group's bidding scheme, which is obtained by subtasks pre-assignment based on the group's internal contract network.

3. The top leader selects the group with the largest bidding value as the bid winner according to the bidding schemes of the group submitted by the group leaders and assigns the target to the group.

4. According to the assigned target, each subset performs tasks according to the corresponding subtask assignment scheme.

5. If a UAV fails to perform tasks due to sudden failure, the tasks shall be assigned to other UAVs in this subset first; if no UAV in the subset can continue to perform, the group leader will release the tasks to other subsets, and then others will determine the pre-assignment scheme based on the contract network within the group. The tendering group shall determine the assistance according to each bidding scheme.

6. If a UAV (including the group leader) detects a sudden target, it shall be reported to the group leader. After receiving the target information, the leader of this subset will release the target to other subsets. Each subset determines the pre-assignment scheme based on the contract network within the group, carries out the bidding based on the inter-group contract network, and, finally, completes the target and subtask assignment.

7. Repeat steps 4–6 until there is no dynamic change.

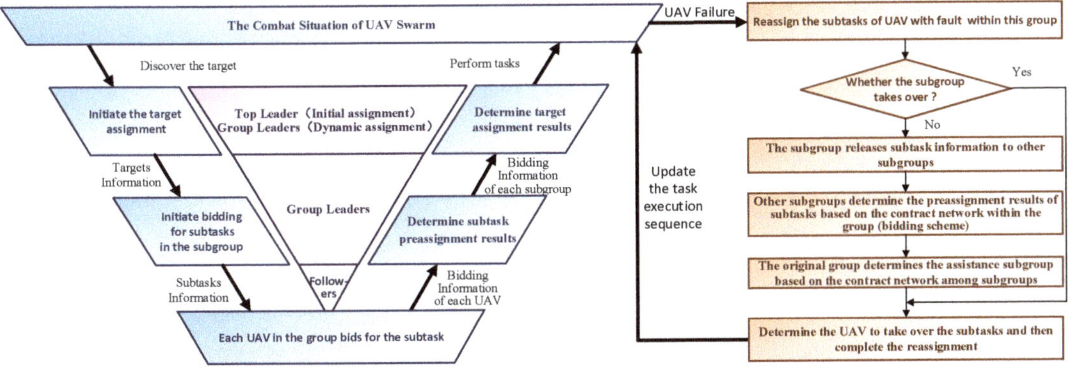

Figure 7. The procedure for cooperative dynamic task assignment.

5. Numerical Simulation

It is assumed that the swarm consists of 1 top leader and 2 subsets, with a total of 11 UAVs, including 2 reconnaissance-attack UAVs, 4 reconnaissance UAVs, and 4 attack UAVs. Among them, reconnaissance-attack UAVs are the group leaders. Consider the UAV swarm communication topology G_0, as shown in Figure 8.

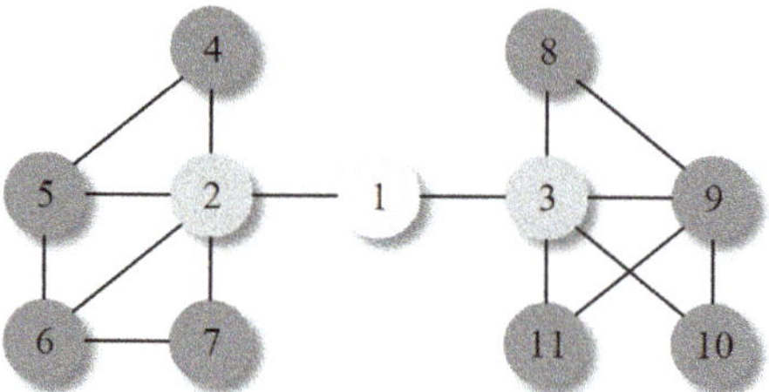

Figure 8. The communication topology G_0 of the UAV swarm.

Each UAV can perform no more than two subtasks, and each subtask is performed jointly by, at most, two UAVs. For flight path preplanning, the UAV kinematics model in reference [35] and the modified artificial potential field method [36–38] are used to realize the air-range estimation considering threat avoidance. The capability parameters of each UAV under the scenario are shown in Table 1, and the parameters of reconnaissance payload and attack payload are shown in Tables 2 and 3, respectively.

Table 1. The table of capability parameters of UAVs.

UAV	Initial Position (m)	The Maximum Air-Range (m)	Fuel Consumption Rate (m^{-1})	Velocity (m/s)	The Maximum Number of Tasks	Mission Capability Reconnaissance	Attack
V1	(2500,20,200)	10,000	0.01	120	2	√	
V2	(0,20,200)	10,000	0.03	80	2	√	√
V3	(5000,20,200)	10,000	0.03	80	2	√	√
V4	(500,20,200)	10,000	0.01	120	2	√	
V5	(1000,20,200)	10,000	0.01	120	2	√	
V6	(1500,20,200)	10,000	0.02	100	2		√
V7	(2000,20,200)	10,000	0.02	100	2		√
V8	(4500,20,200)	10,000	0.01	120	2	√	
V9	(4000,20,200)	10,000	0.01	120	2	√	
V10	(3500,20,200)	10,000	0.02	100	2		√
V11	(3000,20,200)	10,000	0.02	100	2		√

Table 2. The parameter table of UAV reconnaissance payload.

Flight Height H/m	Operating Distance $R_{s,max}$/m	Azimuth Search Angle φ_{max}/°	Pitch Search Angle ϕ_{max}/°	Mounting Angle α/°
200	500	45	30	30

Table 3. The parameter table of UAV attack load.

Flight Velocity V/m	The Minimum Launch Distance $R_{a,min}$/m	The Maximum off-Boresight Angle $\varphi_{a,max}$/°	The Maximum Operating Range of Guidance Equipment d_{max}/m	Maximum Horizontal Detection Angle of Guidance Equipment $\pm \phi_{a,max}$/°	Aiming Time of Guidance Equipment t_a/s
100	80	60	30	±60	0.2

The feasible area of UAV reconnaissance and attack for a target on the ground, based on the above parameter configuration, is shown in Figure 9.

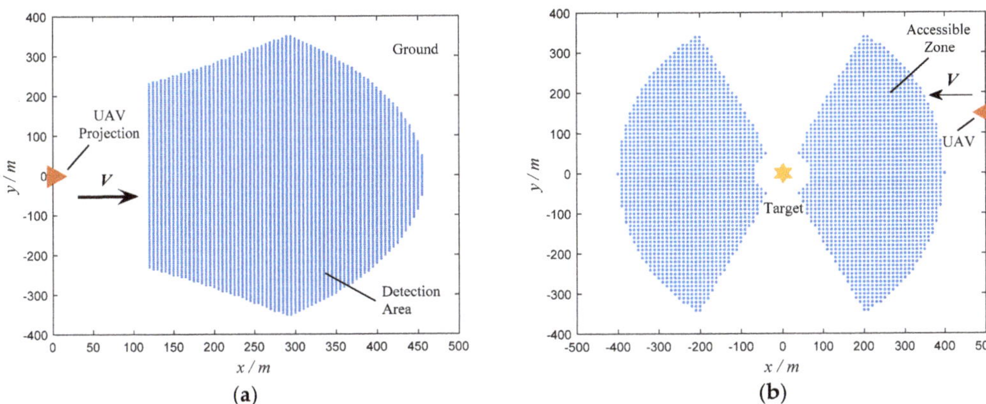

Figure 9. The feasible area of payload: (**a**) the available detection area; (**b**) the available attack zone.

Consider a 5×5 km mission scenario. The information of known targets and the time window constraints of subtasks randomly generated is shown in Table 4.

Table 4. The table of information on known targets.

Target	Coordinate (m)	Task Type		Time Window of S		Time Window of A	
		Reconnaissance S	Attack A	Latest (s)	Earliest (s)	Latest (s)	Earliest (s)
T1	(538.4, 3516.3, 0)	√	√	73.7	93.7	98.7	123.7
T2	(1497.9, 3788.2, 0)	√	√	117.5	137.5	142.5	167.5
T3	(2396.8, 4602.3, 0)	√	√	150.3	170.3	175.3	200.3
T4	(4005.0, 4287.9, 0)	√	√	193.8	213.8	218.8	243.8
T5	(3359.0, 4171.4, 0)	√	√	249.6	269.6	274.6	299.6
T6	(1949.3, 3865.2, 0)	√	√	167.0	187.0	192.0	217.0

5.1. The Initial Task Assignment for Known Targets

Assume that the UAV swarm performs the reconnaissance-and-attack tasks at six enemy targets. The location of each UAV, threat, and target is shown in Figure 10.

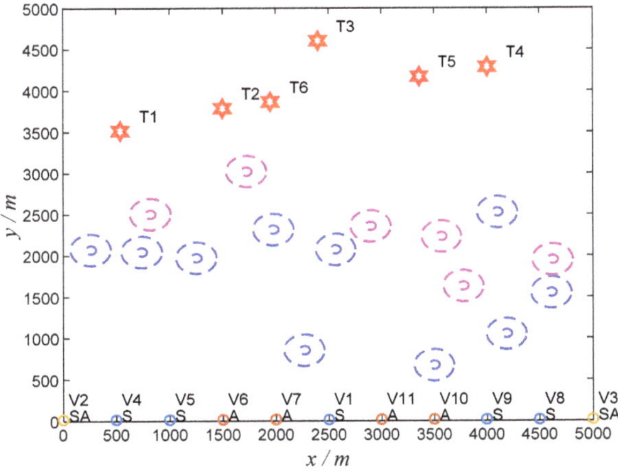

Figure 10. The initial setting of the situation between both sides.

Purple threats represent known threats, and blue threats represent unknown ones, all of which the coordinates are randomly generated. The targets are marked with "T"; Each UAV is marked with "V", where "S" represents the reconnaissance UAV, "A" represents the attack UAV, and "SA" represents the reconnaissance–attack UAV.

In the simulation, the maximum value of the reconnaissance subtask is set at 100, the maximum value of the attack subtask is set at 150, and the attenuation factor λ is set at 0.9.

Adopting the designed UAV swarm cooperative dynamic task assignment approach, the sequential task assignment results for known targets are shown in Table 5. The task execution plan of each UAV is expressed in the format of (Target, Subtask type, Start time (s), Winning bidding value).

Table 5. The task assignment of the UAV swarm.

UAV	Air-Range Estimation (m)	The Number of Tasks	The Execution Planning
V1	0	0	None
V2	3524.00	1	(T1, A, 98.7, 79.52)
V3	0	0	None
V4	3486.00	1	(T1, S, 73.7, 65.14)
V5	4230.00	2	(T2, S, 117.5, 62.14)→(T6, S, 167.0, 59.20)
V6	3756.52	1	(T2, A, 142.5, 74.87)
V7	3837.18	1	(T6, A, 192.0, 73.23)
V8	4298.74	1	(T5, S, 249.6, 57.04)
V9	6475.10	2	(T3, S, 150.3, 51.47)→(T4, S, 193.8, 57.34)
V10	4477.55	1	(T4, A, 218.8, 60.51)
V11	5702.98	2	(T3, A, 175.3, 56.75)→(T5, A, 274.6, 66.90)

According to the known threat information and the unknown threat information detected by reconnaissance UAV during mission execution, the modified artificial potential field method is adopted to avoid local optimization, and the preplanning flight path considering threat avoidance during mission execution can be realized. Combined with the UAV swarm task assignment results, the preplanned path, and the task sequence, the swarm task execution diagram can be obtained, as shown in Figure 11.

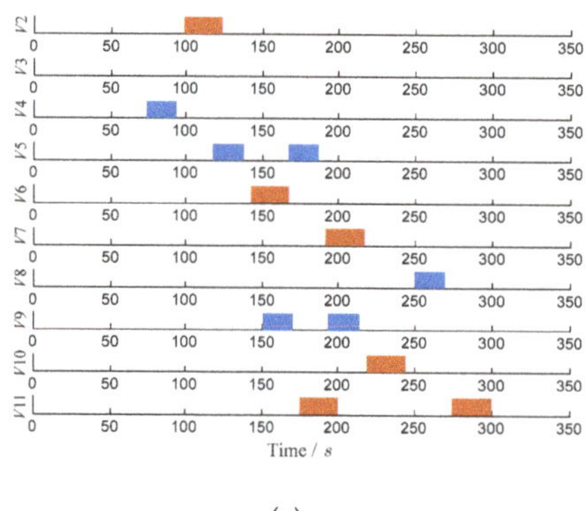

(a)

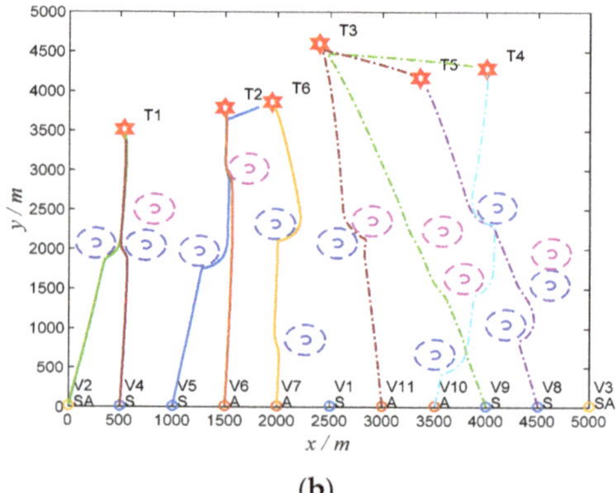

(b)

Figure 11. (**a**) The execution sequence of the UAV swarm; (**b**) the diagram of the UAV swarm performing tasks.

As can be seen in Figure 11, the cooperative task assignment approaches can realize the "reconnaissance–attack" coverage of known targets under the consideration of flight path planning coupling and task time window constraints.

5.2. Dynamic TASK Assignment for Sudden Targets

Considering two unknown targets, T7 and T8, the swarm discovers targets through cooperative detection and then assigns tasks. The information of T7 and T8 is shown in Table 6, and the initial operational situation is shown in Figure 12.

Table 6. The table of information on unknown targets.

Target	Coordinates (m)	Task Type	
		Reconnaissance S	Attack A
T7	(3467.7, 2897.7, 0)	√	√
T8	(1307.7, 3057.7, 0)	√	√

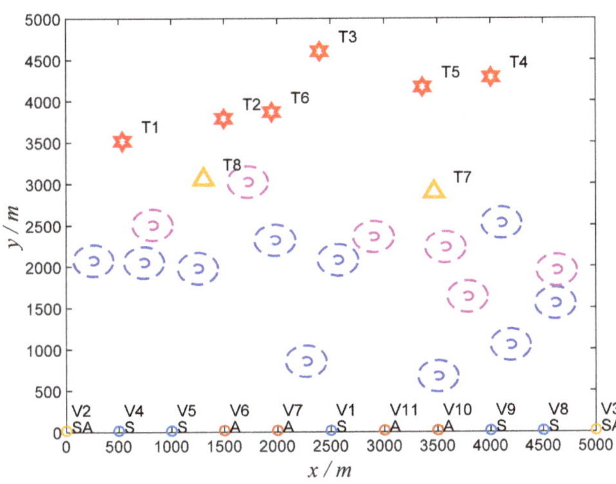

Figure 12. The initial setting of the situation between both sides. (Considering unknowned targets).

The time when unknown targets are discovered by the UAVs and the corresponding subsets that discover the targets are shown in Table 7. The generated subtask time window constraints are shown in Table 7 as well. The sequential task assignment results for the unknown targets are shown in Table 8.

Table 7. Information of unknown targets and corresponding subtasks.

Target	The Time It Is Discovered	Discoverer	Subset to Which the Discoverer Belongs	Time Window of S		Time Window of A	
				Latest (s)	Earliest (s)	Latest (s)	Earliest (s)
T7	237.9	V8 (S)	2	271.0	291.0	291.0	316.0
T8	109.1	V5 (S)	1	155.8	175.8	175.8	200.8

The swarm realizes the coverage of sudden targets through the distributed cooperative dynamic task allocation mechanism after the reconnaissance UAV V8, belonging to Subset 1, and the reconnaissance UAV V5, belonging to Subset 2, detect sudden targets T7 and T8, respectively, during task execution.

The concrete analysis is as follows: The reconnaissance UAV V4 and attack UAV V6, belonging to Subset 1, is assigned to perform the subtasks of sudden target T8 successively in the corresponding time window. The reconnaissance UAV V8 and reconnaissance-attack UAV V3, belonging to Subset 2, successively perform the subtasks of sudden target T7 in the corresponding time window so as to realize the "reconnaissance–strike" coverage of each sudden target.

Table 8. The task assignment of the UAV swarm.

UAV	Air-range Estimation (m)	The Number of Tasks	The Execution Planning
V1	0	0	None
V2	3524.00	1	$(T_1, A, 98.7, 79.52)$
V3	3262.79	1	$(T_7, A, 291.0, 84.77)$
V4	4366.55	2	$(T_1, S, 73.7, 65.14) \rightarrow (T_8, S, 155.8, 32.90)$
V5	4230.00	2	$(T_2, S, 117.5, 62.14) \rightarrow (T_6, S, 167.0, 59.20)$
V6	4502.80	2	$(T_2, A, 142.5, 74.87) \rightarrow (T_8, A, 175.8, 135.13)$
V7	3837.18	1	$(T_6, A, 192.0, 73.23)$
V8	5747.08	2	$(T_5, S, 249.6, 57.04) \rightarrow (T_7, S, 271.0, 27.42)$
V9	6475.10	2	$(T_3, S, 150.3, 51.47) \rightarrow (T_4, S, 193.8, 57.34)$
V10	4477.55	1	$(T_4, A, 218.8, 60.51)$
V11	5702.98	2	$(T_3, A, 175.3, 56.75) \rightarrow (T_5, A, 274.6, 66.90)$

We combine the assignment results with the preplanning flight path and task execution sequence based on the artificial potential field method. The diagrams of UAV swarm task execution and the execution sequence under sudden targets are shown in Figure 13.

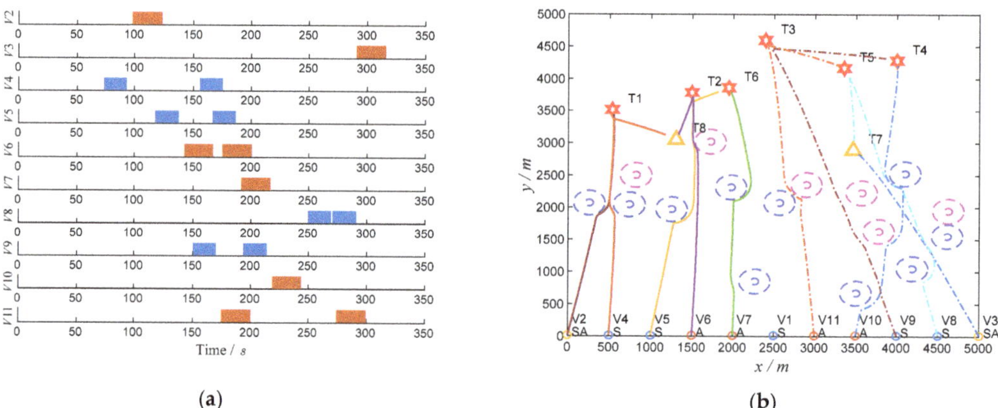

(a)

(b)

Figure 13. (**a**) The execution sequence of the UAV swarm; (**b**) the diagram of the UAV swarm performing tasks. (Considering unknowned targets).

Figure 13 indicates that through the dynamic task assignment approach based on the extended CNP, the swarm can mobilize the UAVs with task capability and the best execution effect obtained after bidding negotiation to complete tasks for each sudden target according to the time window.

5.3. Reassignment for Subtasks of Failed UAV

During the simulation, the reconnaissance UAV V5, belonging to Subset 1, fails and loses the capability to perform tasks. The failure time is 97.5 s.

According to the assignment results in Section 5.1, the reconnaissance UAV V5 prepares to perform the reconnaissance subtask. The fault of V5 triggers dynamic task assignment, and the tasks (T2, s, 117.5, 62.14)→(T6, s, 167.0, 59.20) of V5 become dynamic tasks. The bidding values of each group of potential bidders (including V2, V4, V3, and V8) are shown in Table 9. The reassignment results of failed UAV subtasks obtained by the designed assignment approach are shown in Table 10.

Table 9. The bidding values of potential bidders for dynamic subtasks.

Bidding Value of Subtask	Subset 1		Subset 2	
	V2	**V4**	**V3**	**V8**
T2 (S)	10.72	30.88	−106.49	−48.62
T6 (S)	−6.74	21.83	−2.77	52.89

Table 10. Task assignment under the circumstance of UAV breakdown.

Subtask	Bidding Winner	Winning Bidding Value	Start Time (s)
T2 (S)	V4 (Subset 1)	30.88	117.5 s
T6 (S)	V8 (Subset 2)	52.89	167.0 s

The task assignment results in Table 9 show that UAV V4 of Subset 1 and UAV V8 of Subset 2 replace UAV V5 and continue to perform the tasks, respectively.

Combined with the assignment results, the diagrams of UAV swarm task execution and the execution sequence under UAV V5 failure are shown in Figure 14.

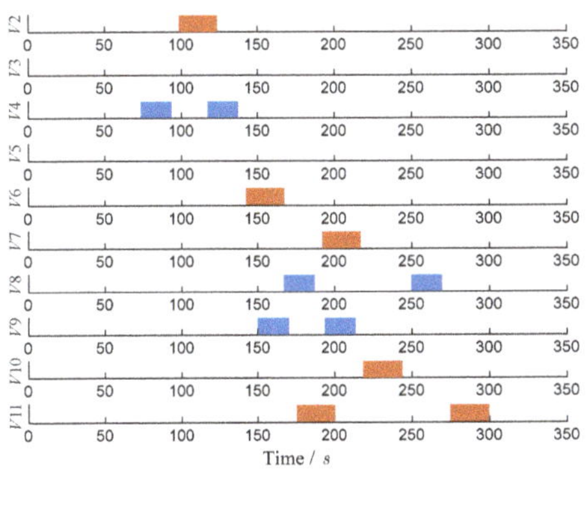

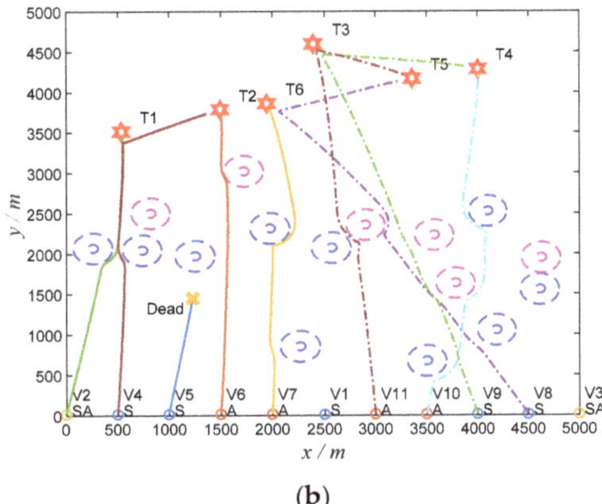

(a)

(b)

Figure 14. (**a**) The execution sequence of the UAV swarm; (**b**) the diagram of the swarm performing tasks under V5 failure.

The explanation of the assignment result is as follows: after the reconnaissance, UAV V5 loses its mission capability, and its subtasks shall be preferentially executed by other reconnaissance UAVs or reconnaissance UAVs (i.e., V2 and V4) in Subset 1. It can be seen from Figure 14a that after the initial assignment in Section 4.1, the number of tasks that can be performed by UAV V4 is 1. Therefore, through the contract network within the group, it can undertake one reconnaissance subtask of V5. At this time, V2 is executing the attack subtask T1(a) according to the initial allocation results. During this execution, it crosses several threat areas, resulting in low survivability estimation, and the bidding value is negative, which is not suitable to continue to perform the subtask. Therefore, the leader UAV V2 of Subset 1 releases the information to Subset 2 for assistance through the contract network among the groups. Subset 2 conducts bidding negotiation through the contract network within the group and finally assigns UAV V8 to assist Subset 1 to take over reconnaissance subtask T6 (s).

5.4. The Analysis of the Real-Time Performance of the Assignment Approach

All the simulation experiments have been implemented on a personal PC; the parameters are Intel Core i5-5350U CPU @ 1.80 GHz 8 GB RAM, and the programming environment is MATLAB 2018b.

5.4.1. Real-Time Performance with Different Problem Scales

The real-time performance of the assignment approach is mainly influenced by three factors: ① the number of subsets of the UAV swarm; ② the number of UAVs in each subset; ③ the number of targets.

To illustrate the impact of the factors above, the variable-controlling method is adopted. The details of simulation cases to be compared are in Table 11. Correspondingly, the comparison of assignment times in different cases is shown in Figure 15.

Table 11. The simulation cases to be compared.

Case	Number of Subsets	Number of UAVs of Each Subset	Number of Targets	The Size of the UAV Swarm
(a)	5, 10, 15, 20, 25	10	10	50, 100, 150, 200, 250
(b)	5	20, 40, 60, 80, 100	10	100, 200, 300, 400, 500
(c)	5	10	5, 10, 15, 20, 25	50

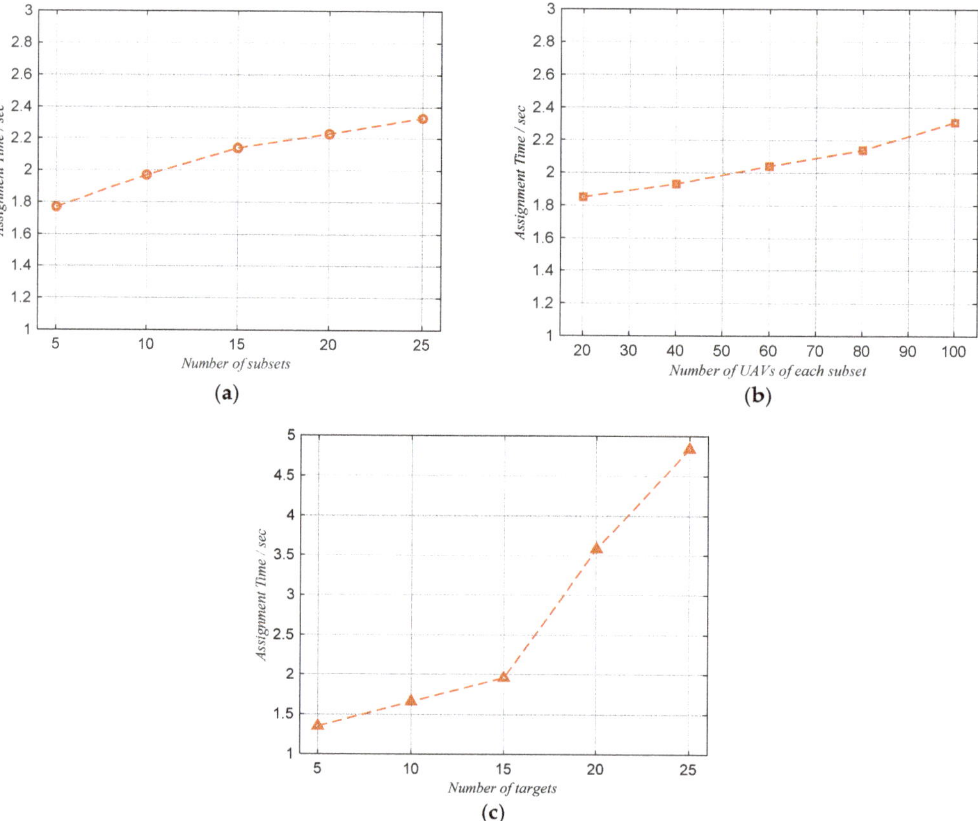

Figure 15. Assignment time varies with (**a**) the number of subsets, (**b**) the number of UAVs of each subset, and (**c**) the number of targets.

According to the comparison among the simulation, on one hand, due to the slowly increasing tendency of the assignment time in Figure 15a,b, it can be concluded that the real-time performance of the proposed approach has low sensitivity with the size of the

UAV swarm. On the other hand, comparing (a) with (b), the task assignment time is more sensitive with the number of subsets than with the number of UAVs of each subset.

Moreover, according to Figure 15c, the task assignment time is sensitive to the number of targets, which indicates that the performance of the approach may become worse with the number of targets increasing.

5.4.2. Algorithm Comparison

In order to verify the effectiveness of the distributed approach based on contract network protocol (CNP) in solving the UAV swarm task assignment problem, it is compared with the centralized task assignment approach used in reference [10]. Scenarios are designed with different sizes of UAV swarm performing multiple task assignments. From the simulation results, both the centralized approach (ACO) and the distributed approach (CNP) are able to obtain the solution to the UAV swarm task assignment problem. However, the distributed approach (CNP) has an obvious advantage in terms of solution efficiency. The solution time for different sized assignment problems is shown in Table 12.

Table 12. The real-time performance between the proposed approach and ACO in reference [10].

The Size of the UAV Swarm	The Number of Targets	The Total Solving Time of the CNP-Based Approach/(Seconds)	The Total Solving Time of ACO (50 Iterations)/(Seconds)
50	10	1.66	41.5
100	10	1.70	59.8
150	10	1.85	74.9
200	10	1.93	96.2
250	10	1.98	112.5
150	15	2.03	88.2
150	20	3.50	124.1
150	25	5.30	179.0

Remark: *Generally, a global optimal solution to such problems can rarely be obtained. Therefore, feasible solutions to the task assignment problem obtained in the solving process are considered in general.*

As a heuristic algorithm, ACO searches in the solution space, learns from the results of each iteration, and finally converges to a feasible solution, which means that ACO requires a lot of computational resources and a much longer time. Simulation results show that this problem becomes more pronounced when the problem size (e.g., size of the UAV swarm, the number of targets) increases.

In contrast to heuristic algorithms such as ACO, CNP merely performs the optimization by bidding on each subset of UAVs as well as UAV individuals without the iterative process of searching for feasible solutions in a large solution space.

The results illustrate that, as a distributed assignment approach, the proposed method in this paper has advantages in real-time performance compared with the ACO proposed in reference [10], which proves the effectiveness of the method to some extent.

6. Conclusions

Aiming at the problem of the cooperative reconnaissance–attack task assignment of UAV swarms in complex environments, this paper proposes a distributed grouping cooperative dynamic task assignment approach by considering multiple constraints, realizes the effective assignment of cooperative reconnaissance–attack tasks to multiple targets, and optimizes the combat effectiveness of the swarm. The main conclusions include:

(1) The proposed extended CNP algorithm, which is based on the determination mechanism of cooperators and the selection mechanism of sequential tasks, with the bidding function considering the constraints of sequence, flight path, and threat; it can

realize the reasonable assignment of reconnaissance and attack tasks on multiple targets under multiple constraints and the optimization of UAV swarm task execution efficiency.

(2) The proposed dynamic task assignment strategy based on the event-trigger mechanism constructs the overall architecture of cooperative dynamic task assignment for the distributed grouping of the UAV swarm and improves the adaptability of the swarm to the dynamic environment and sudden targets during task execution.

(3) Three typical simulation scenarios are designed. The simulation results show that the task assignment approach in this paper can solve the problem of cooperative sequential dynamic task assignment when the UAV swarm with subsets performs reconnaissance–attack tasks in a complex environment with incomplete situational awareness and sudden targets and realize reconnaissance and attack coverage on each target. Moreover, the real-time performance of the assignment approach has been analyzed, which indicates that the proposed approach has low sensitivity to the size of the UAV swarm.

Author Contributions: Conceptualization, B.Q. and D.Z.; methodology, B.Q.; software, B.Q.; validation, B.Q.; formal analysis, B.Q.; investigation, B.Q. and D.Z.; resources, D.Z. and S.T.; data curation, B.Q. and M.W.; writing—original draft preparation, B.Q.; writing—review and editing, B.Q.; visualization, B.Q.; supervision, D.Z. and S.T.; project administration, D.Z. and S.T; funding acquisition, D.Z. All authors have read and agreed to the published version of the manuscript.

Funding: The research was funded by the National Natural Science Foundation of China, grant number (61933010) and (61903301).

Institutional Review Board Statement: Not applicable.

Informed Consent Statement: Not applicable.

Data Availability Statement: The data presented in this study are available upon request from the corresponding author.

Conflicts of Interest: The authors declare no conflict of interest.

References

1. Zhang, Y.F.; Lin, D.F.; Zheng, D.; Cheng, Z.H.; Tang, P. Task Allocation and Trajectory Optimization of UAV for Multi-target Time-space Synchronization Cooperative Attack. *Acta Armamentarii* **2021**, *42*, 1482–1495. (In Chinese)
2. Jiang, X.; Zeng, X.L.; Sun, J.; Chen, J. Research status and prospect of distributed optimization for multiple aircraft. *Acta Aeronaut. Astronaut. Sin.* **2021**, *42*, 90–105. (In Chinese)
3. Zhang, J.; Xing, J.H. Cooperative task assignment of multi-UAV system. *Chin. J. Aeronaut.* **2020**, *33*, 2825–2827. [CrossRef]
4. Duan, H.B.; Zhao, J.X.; Deng, Y.M.; Shi, Y.H. Dynamic Discrete Pigeon-Inspired Optimization for Multi-UAV Cooperative Search-Attack Mission Planning. *IEEE Trans. Aerosp. Electron. Syst.* **2020**, *57*, 706–720. [CrossRef]
5. Ye, F.; Chen, J.; Tian, Y.; Jiang, T. Cooperative Task Assignment of a Heterogeneous Multi-UAV System Using an Adaptive Genetic Algorithm. *Electronics* **2020**, *9*, 687. [CrossRef]
6. Jia, Z.Y.; Yu, J.Q.; Ai, X.L.; Xu, X.; Yang, D. Cooperative multiple task assignment problem with stochastic velocities and time windows for heterogeneous unmanned aerial vehicles using a genetic algorithm. *Aerosp. Sci. Technol.* **2018**, *76*, 112–125. [CrossRef]
7. Wu, W.N.; Wang, X.G.; Cui, N.G. Fast and coupled solution for cooperative mission planning of multiple heterogeneous unmanned aerial vehicles. *Aerosp. Sci. Technol.* **2018**, *79*, 131–144. [CrossRef]
8. Wang, Z.; Liu, L.; Long, T.; Wen, Y.L. Multi-UAV recon-naissance task allocation for heterogeneous targets using an opposition-based genetic algorithm with double chromosome encoding. *Chin. J. Aeronaut.* **2018**, *31*, 339–350. [CrossRef]
9. Wei, C.; Ji, Z.; Cai, B. Particle Swarm Optimization for Cooperative Multi-Robot Task Allocation: A Multi-Objective Approach. *IEEE Robot. Autom. Lett.* **2020**, *5*, 2530–2537. [CrossRef]
10. Su, F.; Chen, Y.; Shen, L.C. UAV Cooperative Multi-task Assignment Based on Ant Colony Algorithm. *Acta Aeronaut. Astronaut. Sin.* **2008**, *29*, 184–191. (In Chinese)
11. Zhen, Z.; Xing, D.; Gao, C. Cooperative search-attack mission planning for multi-UAV based on intelligent self-organized algorithm. *Aerosp. Sci. Technol.* **2018**, *76*, 402–411. [CrossRef]
12. Chen, Y.B.; Yang, D.; Yu, J.Q. Multi-UAV Task Assignment with Parameter and Time-Sensitive Uncertainties Using Modified Two-Part Wolf Pack Search Algorithm. *IEEE Trans. Aerosp. Electron. Syst.* **2018**, *54*, 2853–2872. [CrossRef]
13. Gerkey, B.P.; Mataric, M.J. Sold!: Auction methods for multirobot coordination. *IEEE Trans. Robot. Autom.* **2002**, *18*, 758–768. [CrossRef]
14. Choi, H.L.; Brunet, L.; How, J.P. Consensus-Based De-8kcentralized Auctions for Robust Task Allocation. *IEEE Trans. Robot.* **2009**, *25*, 912–926. [CrossRef]

15. Luo, L.Z.; Chakraborty, N.; Sycara, K. Distributed Algorithms for Multirobot Task Assignment with Task Deadline Constraints. *IEEE Trans. Autom. Sci. Eng.* **2015**, *12*, 876–888. [CrossRef]

16. Farinelli, A.; Iocchi, L.; Nardi, D. Distributed on-line dynamic task assignment for multi-robot patrolling. *Auton. Robot.* **2016**, *41*, 1–25. [CrossRef]

17. Oh, G.; Kim, Y.; Ahn, J.; Choi, H.L. Market-Based Task Assignment for Cooperative Timing Missions in Dynamic Environments. *J. Intell. Robot. Syst.* **2017**, *87*, 1–27. [CrossRef]

18. Moore, B.J.; Passino, K.M. Distributed Task Assignment for Mobile Agents. *IEEE Trans. Autom. Control.* **2007**, *52*, 749–753. [CrossRef]

19. Yang, H.Z.; Wang, Q. A multi-AUV dynamic task allocation method based on ant colony labor division model. *Control. Decis.* **2021**, *36*, 1911–1919. (In Chinese)

20. Li, X.M.; TANG, J.Y.; DAI, J.J.; Bo, N. Dynamic Coalition Task Allocation of Heterogeneous Multiple Agents. *J. Northwestern Polytech. Univ.* **2020**, *38*, 1094–1104. (In Chinese) [CrossRef]

21. Ni, J.J.; Tang, M.; Chen, Y.N.; Cao, W.D. An Improved Cooperative Control Method for Hybrid Unmanned Aerial-Ground System in Multitasks. *Int. J. Aerosp. Eng.* **2020**, *2020*, 1–14. [CrossRef]

22. Yao, W.R.; Qi, N.M.; Wan, N.; Liu, Y.B. An iterative strategy for task assignment and path planning of distributed multiple unmanned aerial vehicles. *Aerosp. Sci. Technol.* **2019**, *86*, 455–464. [CrossRef]

23. Zhan, C.; Zeng, Y. Aerial–Ground Cost Tradeoff for Multi-UAV-Enabled Data Collection in Wireless Sensor Networks. *IEEE Trans. Commun.* **2020**, *68*, 1937–1950. [CrossRef]

24. Wang, R.R.; Wei, W.L.; Yang, M.C.; Liu, W. Task allocation of multiple UAVs considering cooperative route planning. *Acta Aeronaut. Et Astronaut. Sin.* **2020**, *41*, 24–35. (In Chinese)

25. Ismail, A.; Bagula, B.A.; Tuyishimire, E. Internet-Of-Things in Motion: A UAV Coalition Model for Remote Sensing in Smart Cities. *Sensors* **2018**, *18*, 2184. [CrossRef]

26. Fu, X.W.; Feng, P.; Gao, X. Swarm UAVs Task and Resource Dynamic Assignment Algorithm Based on Task Sequence Mechanism. *IEEE Access* **2019**. [CrossRef]

27. Yan, F.; Zhu, X.P.; Zhou, Z.; Tang, Y. Real-time task allocation for a heterogeneous multi-UAV simultaneous attack. *SCIENTIA SINICA Inf.* **2019**, *49*, 555–569. (In Chinese) [CrossRef]

28. Chen, P.; Yan, F.; Liu, Z.; Cheng, G.D. Heterogeneous unmanned aerial vehicles task allocation with communication constraints. *Acta Aeronaut. Et Astronaut. Sin.* **2021**, *42*, 525844. (In Chinese)

29. Jia, G.W.; Wang, J.F. Research review of UAV swarm mission planning method. *Syst. Eng. Electron.* **2021**, *43*, 99–111. (In Chinese)

30. Wu, Q.B.; Zhou, S.L.; Liu, W.; Yin, G.Y. Multi-unmanned aerial vehicles cooperative search based on central-distributed model predictive control. *Control. Theory Appl.* **2015**, *32*, 1414–1421. (In Chinese)

31. Tian, L.; Wang, M.Y.; Zhao, Q.L.; Wang, X.D.; Song, X.; Ren, Z. Distributed time-varying group formation tracking for cluster systems under switching topologies. *Scientia Sinica Inf.* **2020**, *50*, 408–423. (In Chinese)

32. Dong, X.W.; Li, Q.D.; Zhao, Q.L.; Ren, Z. Time-varying group formation analysis and design for second-order multi-agent systems with directed topologies. *Neurocomputing* **2016**, *205*, 367–374. [CrossRef]

33. Su, F. Research on Distributed Online Cooperative Mission Planning for Multiple Unmanned Combat Aerial Vehicles in Dynamic Environment. Ph.D. Thesis, National University of Defense Technology, Changsha, China, 2013. (In Chinese).

34. Ma, D.L.; Chen, Y. Assessments of air-to-surface operational effectiveness for aircraft during combat sortie. *J. Beijing Univ. Aeronaut. Astronaut.* **2000**, *26*, 1413–1417. (In Chinese)

35. Wang, Y.X.; Zhang, T.; Cai, Z.H.; Zhao, J.; Wu, K. Multi-UAV coordination control by chaotic grey wolf optimization based distributed MPC with event-triggered strategy. *Chin. J. Aeronaut.* **2020**, *33*, 2877–2897. [CrossRef]

36. Zhu, Y.; Zhang, T.; Song, J.Y. Study on the Local Minima Problem of Path Planning Using Potential Field Method in Unknown Environments. *Acta Autom. Sin.* **2010**, *36*, 1122–1130. [CrossRef]

37. Fan, S.P.; Qi, Q.; Lu, K.F.; Wu, G.; Li, L. Autonomous Collision Avoidance Technique of Cruise Missiles Based on Modified Artificial Potential Method. *Trans. Beijing Inst. Technol.* **2018**, *38*, 828–834.

38. Ge, S.S.; Cui, Y.J. New potential functions for mobile robot path planning. *IEEE Trans. Robot. Autom.* **2000**, *16*, 615–620. [CrossRef]

applied sciences

MDPI

Article

UAV-Cooperative Penetration Dynamic-Tracking Interceptor Method Based on DDPG

Yuxie Luo [1], Jia Song [1,*], Kai Zhao [1] and Yang Liu [2]

[1] School of Astronautics, Beihang University (BUAA), Beijing 100191, China; luoyuxie@buaa.edu.cn (Y.L.); zk19970207@buaa.edu.cn (K.Z.)

[2] The Seventh Research Division, Beihang University (BUAA), Beijing 100191, China; ylbuaa@163.com

* Correspondence: songjia@buaa.edu.cn

Abstract: The multi-UAV system has stronger robustness and better stability in combat. Therefore, the collaborative penetration of UAVs has been extensively studied in recent years. Compared with general static combat scenes, the dynamic tracking and interception of equipment penetration are more difficult to achieve. To realize the coordinated penetration of the dynamic-tracking interceptor by the multi-UAV system, the intelligent UAV model is established by using the deep deterministic policy-gradient algorithm, and the reward function is constructed using the cooperative parameters of multiple UAVs to guide the UAV to proceed with collaborative penetration. The simulation experiment proved that the UAV finally evaded the dynamic-tracking interceptor, and multiple UAVs reached the target at the same time, realizing the time coordination of the multi-UAV system.

Keywords: multi-UAV; deep deterministic policy gradient; cooperative penetration; dynamic-tracking-interceptor component

Citation: Luo, Y.; Song, J.; Zhao, K.; Liu, Y. UAV-Cooperative Penetration Dynamic-Tracking Interceptor Method Based on DDPG. *Appl. Sci.* **2022**, *12*, 1618. https://doi.org/10.3390/app12031618

Academic Editors: Wei Huang and Dong Zhang

Received: 24 December 2021
Accepted: 1 February 2022
Published: 3 February 2022

Publisher's Note: MDPI stays neutral with regard to jurisdictional claims in published maps and institutional affiliations.

1. Introduction

Compared with traditional manned aerial vehicles, unmanned aerial vehicles (UAVs) can be autonomously controlled or remotely controlled, which have the advantages of low requirements on the combat environment and strong battlefield survivability, and they can be used to perform a variety of complex tasks [1,2]. Therefore, UAVs have been widely studied and applied [3]. With the continuous application of UAVs in the military field, the system composed of a single UAV has gradually revealed the problems of poor flexibility and low stability [4,5]. The cooperative combat method of using a multi-UAV system composed of multiple UAVs has become a new main research direction [6,7]. Under the conditions of the modernized and networked battlefield, the air cluster composed of multiple UAVs has the air power to continuously launch the required strikes, forcing the enemy to spend more resources and deal with more fighters, thereby enhancing the overall capability and overall performance of military combat confrontation.

Multi-UAV-cooperative penetration is one of the key issues to achieve multi-UAV-cooperative combat. Multiple UAVs start from the same location or different locations, and finally arrive at the same place. At present, UAVs' penetration-trajectory-planning methods mainly include the A* algorithm [8], the artificial potential field method [9,10], and the RRT algorithm [11]. Most of the application scenarios of these methods are environments with static no-fly zones and rarely consider dynamic threats. The A* algorithm is a typical grid method. This type of method rasterizes the map for planning, but the size of the grid will have a greater impact on the result and is difficult to determine. Based on the artificial potential field law, it is easy to fall into the local optimum, leading to the unreachable target. When there is a dynamic-tracking interceptor in the environment, the environment information becomes complicated, and real-time planning requirements are put forward for UAVs. Therefore, traditional algorithms cannot meet the requirements.

Multi-UAV-cooperative penetration generally requires that multiple UAVs achieve time coordination to penetrate defenses according to different trajectories and finally reach the target area at the same time or according to a certain time sequence. When the UAVs depart from different locations, it will greatly increase the difficulty of collaboration. The multi-UAV-coordination algorithm can be improved based on the traditional single-UAV-penetration algorithm adapted to the multi-UAV environment. Chen uses the optimal-control method to improve the artificial potential-field method to achieve multi-UAV coordination [11,12], and Kothari improves the RRT algorithm to achieve multi-UAV coordination [13]. At the same time, there are a large number of methods for cooperative control of UAVs based on the graph theory [3]. Li proposed a multi-UAV-collaboration method based on the graph theory [14]. Ruan proposed a multi-UAV-coordination method based on multi objective optimization [15]. The above methods all realize the coordination of multiple UAVs, but their algorithms lack the research on dynamic environments and cannot adapt to the complex and dynamic battlefield environment. Aimed at the environment with dynamic threats, this paper proposes a method based on deep reinforcement learning to achieve multi-UAV-cooperative penetration.

Reinforcement learning is an important branch of machine learning. Its main feature is to evaluate the action policy of the agent based on the final rewards and through the interaction and trial and error between the agent and the environment. Reinforcement learning is a far-sighted machine-learning method that considers long-term rewards [16,17]. It is often used to solve sequential decision-making problems. Reinforcement learning can not only be applied in a static environment but also when the parameters of the environment are constantly changing, and the agent can also be applied in a dynamic environment [16–18]. The research of reinforcement learning is mostly concentrated in the field of a single agent, but there is also a large body of research on reinforcement-learning algorithms for multiagent systems. There is also much research on applying reinforcement learning to UAVs. Pham successfully applied deep reinforcement learning to UAVs to realize autonomous navigation of UAVs [19]. Wang also studied the autonomous navigation of UAVs in large, unknown, and complex environments based on deep reinforcement learning [20]. Wang applied reinforcement learning to the target search and tracking of UAVs [21]. Based on deep reinforcement learning, Yang studied the task scheduling problem of UAV clusters and solved the application problem of reinforcement learning in multi-UAV systems [22]. Through the investigation of relevant literature, it can be found that the application of deep reinforcement learning to multi-UAV systems is a feasible method, which can be used to achieve complex multi-UAV-system tasks and has great research potential.

The main work of this paper uses the multi-UAV-cooperative penetration dynamic-tracking interceptor as the scenario. Based on the deep reinforcement-learning DDPG algorithm, we establish the intelligent UAV model and realize the multi-UAV-cooperative penetration dynamic-tracking interceptor by designing the reward function related to coordination and penetration. The simulation experiment results show that the trained multi-UAV system can achieve cooperative attack tasks from different initial locations, which proves the application potential of artificial intelligence methods, such as reinforcement learning in the implementation of coordinated tasks in UAV clusters.

2. Problem Statement

2.1. Motion Scene

This paper solves the problem of multi-UAV-cooperative penetration of dynamic interceptors. We assume multiple UAVs are respectively numbered as $L = \{1, 2, \ldots, n\}$. The scene is set as a two-dimensional engagement plane. Figure 1 is the schematic diagram of the motion scene of the multi-UAVs.

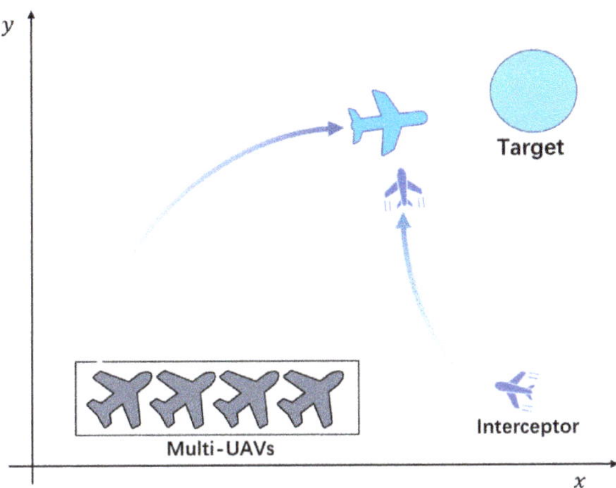

Figure 1. The motion scene of the multi-UAVs.

Based on reinforcement learning algorithms, the following are the requirements:

- The UAV is not intercepted by interceptors when it is moving;
- Multi-UAVs finally reach the target area at the same time.

2.2. UAV Movement Model

First, a two-dimensional movement model of the UAV is established. Figure 2 is the schematic diagram of the movement of the UAV. It is assumed that the linear velocity of the UAV is constant, and the angular velocity, ω, is a continuously variable value. It is assumed that the angle between the movement direction of the UAV and the x-axis is the azimuth angle, θ.

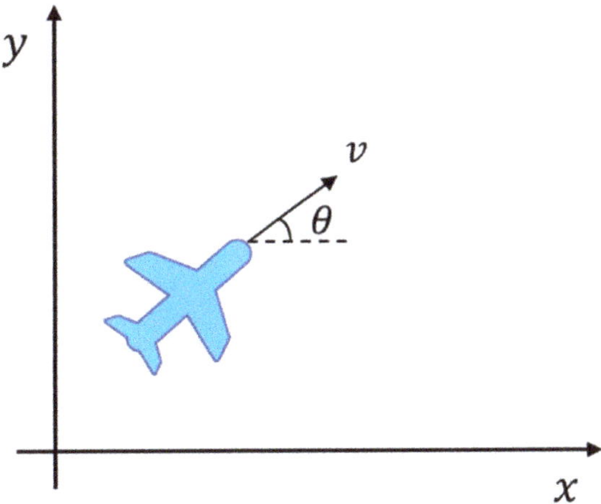

Figure 2. Schematic diagram of UAV movement.

The movement of the UAV is divided into x and y directions. First, the current azimuth angle is obtained by integrating the angular velocity of the UAV, and then the velocity is decomposed on the coordinate axis by the azimuth angle, θ, and finally, the position information of the UAV is obtained through integration. The mathematical model is shown in (1):

$$\begin{cases} \theta = \theta_0 + \int_0^t \omega dt \\ v_x = v \sin \theta \\ v_y = v \cos \theta \\ x = x_0 + \int_0^t v_x dt \\ y = y_0 + \int_0^t v_y dt \end{cases} \tag{1}$$

2.3. Dynamic-Interceptor Design

Compared with the static environment, the position of the dynamic-interceptor changes in real time in the environment, so the UAV is required to be able to perform real-time planning. In this paper, the dynamic interceptor is defined as a tracking interceptor according to the proportional-guided pursuit law. Compared with most common dynamic interceptors with simple motion rules, such as interceptors that cycle in a uniform linear motion or circular motion, the tracking interceptor has stronger uncertainty, and it is difficult to evade by predicting the motion. The movement requirements of the multi-UAV system are much higher. In this paper, it is assumed that the linear velocity of the tracking interceptor is constant, and the angular velocity is calculated by proportional guidance. The schematic diagram of its movement is shown in Figure 3.

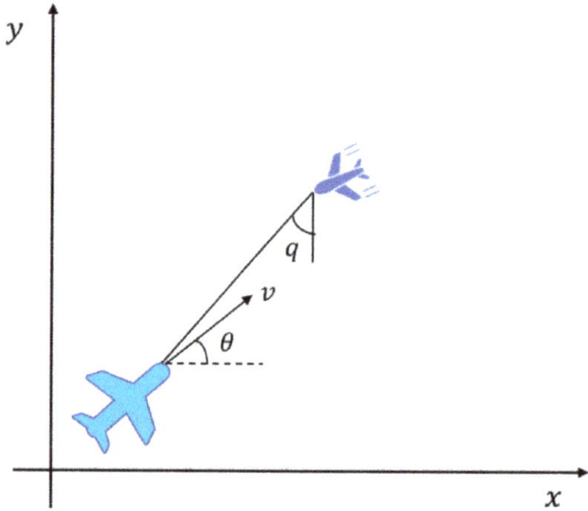

Figure 3. Schematic diagram of interceptor's movement.

The basic principle of the proportional-guidance method is to make the interceptor's rotational angular velocity proportional to the line-of-sight angular velocity. Next, the proportional-guidance mathematical model of the interceptor is introduced.

Assuming that the position of the interceptor is x_b, y_b, the speed is v_{x_b}, v_{y_b}, the position of the UAV is x, y, and the speed is v_x, v_y, and the relative position and speed of the interceptor to the UAV are shown in (2):

$$\begin{cases} x_r = x_b - x \\ y_r = y_b - x \\ v_{x_r} = v_{x_b} - v_x \\ v_{y_r} = v_{y_b} - v_y \end{cases} \tag{2}$$

The interceptor's line of sight angle to the UAV can be obtained as:

$$q = \arctan\left(\frac{x_r}{y_r}\right) \tag{3}$$

To obtain the line-of-sight angular velocity, (3) is derived, as shown in (4):

$$\dot{q} = \frac{v_{y_r} x_r - v_{x_r} y_r}{x_r^2 + y_r^2} \tag{4}$$

The rotational angular velocity of the dynamic-tracking interceptor can be obtained by (5):

$$\omega_b = K\dot{q} \tag{5}$$

K is the proportional guidance coefficient, taking $K = 2$.

Based on the angular velocity of rotation calculated by the proportional guidance, the interceptor performs a two-dimensional movement according to the angular velocity of the rotation obtained by the proportional guidance. The movement model is similar to that of a UAV:

$$\begin{cases} \theta_b = \theta_{b0} + \int_0^t \omega_b dt \\ v_{bx} = v_b \sin\theta_b \\ v_{by} = v_b \cos\theta_b \\ x_b = x_{b0} + \int_0^t v_{bx} dt \\ y_b = y_{b0} + \int_0^t v_{by} dt \end{cases} \tag{6}$$

3. Deep Deterministic Policy-Gradient Algorithm

The DDPG algorithm is a branch of reinforcement learning [5]. The basic process of reinforcement-learning training is that the agent performs an action based on the current observation state. This action acts on the agent's training environment and returns a reward and a new state observation. The goal of training is to maximize the final reward. Reinforcement learning does not need to give any artificial strategies and guidance during training but only needs to give the reward function when the environment is in various states. This is also the only part of training that can be adjusted artificially. Figure 4 shows the basic process of reinforcement learning.

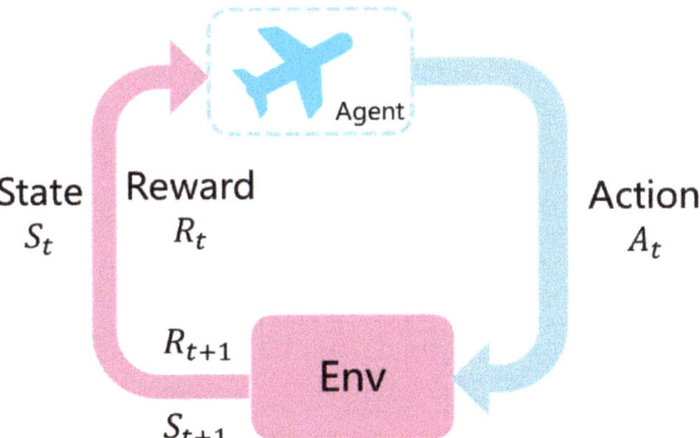

Figure 4. The basic process of reinforcement learning.

The DDPG algorithm is an actor-critic framework algorithm that solves the problem of applying reinforcement learning in continuous space. There are two networks in the DDPG algorithm, namely, the state-action value function network $Q(s, a|\theta^Q)$ using θ^Q parameters and the policy network $\mu(s|\theta^\mu)$ using θ^μ parameters. At the same time, two concepts are introduced, target network and experience replay. When the value function network is updated, the current value function is used to fit the future state-action value function. If both state-action value functions use the same network, it is difficult to fit during training. Therefore, the concept of the target network is introduced. The target network is used as the future state-action value function, which is the same as the state-action value function network to be updated, except that it is not updated in real time but is updated according to the state-action value function network when the state-action value function network is updated to a certain extent. The policy network also adopts the same training method in DDPG. The experience replay is a function of storing state transfer (s_t, a_t, r_t, s_{t+1}), and it will be stored in the experience replay pool every time the agent performs an action that causes the state to transfer. When the value function is updated, it will not be updated directly according to the action of the current policy, but the state transition value will be extracted from the experience playback pool for updating. The advantage of such an update is that the training and learning of the network is more efficient.

Before training, the value function network, $Q(s, a|\theta^Q)$, and the policy network, $\mu(s|\theta^\mu)$, are first randomly initialized, and then the target network, Q' and μ', are initialized. It is also necessary to initialize an action random noise, $\mathcal{N}$, which is conducive to the agent's exploration. During training, the agent selects and executes actions, $a_t = \mu(s_t \mid \theta^\mu) + \mathcal{N}_t$, based on the current policy network and action noise and receives rewards, r_t, and new state observations, s_{t+1}, based on environment feedback. The state transition (s_t, a_t, r_t, s_{t+1}) is stored in the experience replay pool. After that, N state transitions (s_i, a_i, r_i, s_{i+1}) are randomly selected to update the value function network. The principle of updating the value function network is to minimize the loss function. The mathematical expression of the loss function is as follows:

$$L = \frac{1}{N} \sum_i \left(y_i - Q\left(s_i, a_i \mid \theta^Q \right) \right)^2 \tag{7}$$

where $y_i = r_i + \gamma Q'\left(s_{i+1}, \mu'\left(s_{i+1} \mid \theta^{\mu'} \right) \mid \theta^{Q'} \right)$, y_i is only related to θ^Q.

Assuming the objective function of training is the following:

$$J(\theta^\mu) = E_{\theta^\mu}\left[r_1 + \gamma r_2 + \gamma^2 r_3 + \ldots \gamma^N r_N \right] \tag{8}$$

where γ is the discount factor. The policy network is updated according to the gradient of the objective function, and its mathematical expression is as follows:

$$\nabla_{\theta^\mu} J \approx \frac{1}{N} \sum_i \nabla_a Q\left(s, a \mid \theta^Q \right) \Big|_{s=s_i, a=\mu(s_i)} \nabla_{\theta^\mu} \mu(s \mid \theta^\mu) \Big|_{s_i} \tag{9}$$

After training and finally updating the target network, the mathematical expression is as follows:

$$\begin{cases} \theta^{Q'} \leftarrow \tau\theta^Q + (1-\tau)\theta^{Q'} \\ \theta^{\mu'} \leftarrow \tau\theta^\mu + (1-\tau)\theta^{\mu'} \end{cases} \tag{10}$$

To extend DDPG to a multi-UAV system, multiple actors and a critic must exist in the system. During each training, the value function network evaluates the policies of all UAVs in the environment, and the UAVs update their respective policy networks based on the evaluation and independently choose to execute actions. Figure 5 shows the structure of the multi-UAV DDPG algorithm.

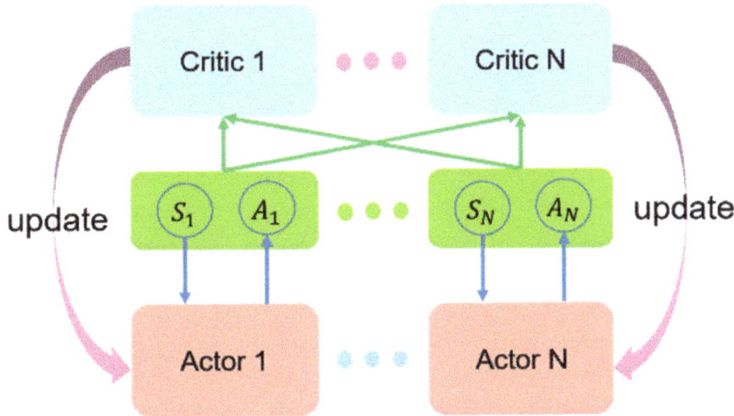

Figure 5. Multi-UAV DDPG algorithm.

The algorithm design also needs to construct the UAV's action space, state space, reward function, and termination conditions. In this paper, all UAVs are tested and simulated using small UAV models for the purpose of preliminary verification of the algorithm.

3.1. Action-Space Design

It can be seen from the motion model of the UAV that the action the UAV can perform is to change the angular velocity so the action space of multiple UAVs is designed as $A = \{\omega_1, \omega_2, \ldots, \omega_n\}$, which is the collection of the angular velocity of multiple UAVs.

3.2. State-Space Design

To realize the coordinated penetration of UAVs, the design of the state space should include the UAVs' positions, x_i, y_i, speed, v_{x_i}, v_{y_i}, and the central position of the target area, x_t, y_t. At the same time, it is necessary to introduce the state observation of the interceptor position, x_b, y_b. Therefore, the state space is set to $S = \{x_i, y_i, v_{x_i}, v_{y_i}, x_t, y_t, x_b, y_b\}, i \in L$.

3.3. Termination Conditions Design

The termination conditions are divided into four items, namely, out of bounds, collision, timeout, and successful arrival.

a) **Out of bounds**: When the movement of the UAV exceeds the environmental boundary, it will be regarded as a mission failure; an end signal and a failure signal will be given.

b) **Collision**: When the UAV is captured by the interceptor, that is, the distance between the two, $d_b = \sqrt{(x - x_b)^2 + (y - y_b)^2} \leq d_{collision}$, it is regarded as a mission failure; an end signal and a failure signal are given.

c) **Timeout**: When the training time exceeds the maximum exercise time, the task will be regarded as a failure; an end signal and a failure signal will be given.

d) **Successful arrival**: When the UAV successfully reaches the target area, the mission is successful; an end signal and a success signal are given.

When any UAV in the environment finishes training, all UAVs finish training and give a failure or success signal according to the distance from the target point.

3.4. Reward-Function Design

The sparse reward problem is a common problem when designing the reward function. This problem will affect the training process of the UAV, prolong the training time of the UAV and even fail to achieve the training goal. To better achieve collaborative tasks and solve the problem of sparse reward, the reward-function design is divided into four parts, namely, the distance-reward function, R_d, which is related to the distance between the UAV and the target, the cooperative reward, R_{co}, which is used to constrain the position of the UAV, the mission success reward, R_s, and the mission failure reward, R_{fail}. The reward function is linearized to improve the efficiency of UAV cluster training. One of the UAVs is an example to introduce the design of the reward function.

Assume $d_i = \sqrt{(x_i - x_t)^2 + (y_i - y_t)^2}, i \in L$ represents the distance between one UAV and the target area, d_{target}, representing the distance between the UAV's initial location and the target area.

The distance reward, R_d, is related to the distance from the UAV to the target area. The closer the distance, the greater the distance-reward value. This type of reward is the key reward on whether the UAV can reach the target area. The specific form is shown in (11):

$$R_d = 0.6 - d_i / d_{target} \tag{11}$$

The cooperative reward is related to the cooperative parameters in the UAV cluster. Here, the difference between the farthest and closest distance between the UAV and the target area is selected as the coordination parameter. Its specific expression is shown in (12). When there are two UAVs in the cluster, the distribution diagram is shown in Figure 6.

$$R_{co} = R_d \times \left(1 - (d_{max} - d_{min}) / d_{target}\right) \tag{12}$$

where $d_{max} = \{d_1, d_2, \ldots, d_n\}_{max}$, $d_{min} = \{d_1, d_2, \ldots, d_n\}_{min}$, respectively, represent the maximum and minimum distances between the UAV and the target area in the environment. It can be seen from the mathematical expression and distribution diagram that there are two major distribution trends for synergistic rewards. When the maximum-distance difference of the UAV cluster is smaller, its value is larger, which will lead the UAVs to move towards time coordination. When the UAV is closer to the target area, its value is larger, and the UAV will be guided to reach the target area.

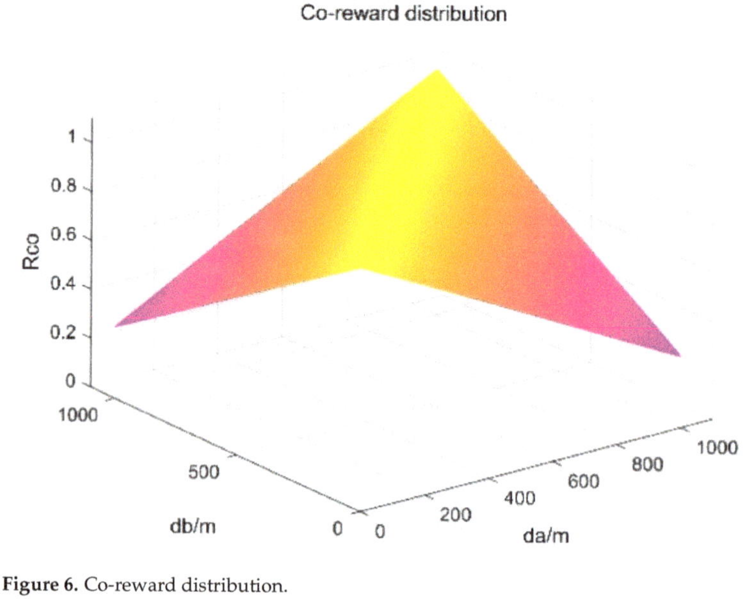

Figure 6. Co-reward distribution.

When the UAV receives the success signal and finally reaches the target area, the mission is successful, and it will give a success reward. The success reward is also related to the farthest distance between the UAV cluster and the target point. The closer the distance, the greater the success reward, as shown in (13).

$$R_{fail} = -1000 \tag{13}$$

When the final mission of the UAV fails, that is, when it is intercepted by an interceptor and fails to reach the target area, it will give a negative reward of failure, as shown in (14).

$$R_s = 250 + 2500/d_{max} \tag{14}$$

By linearly superimposing the above multiple reward functions, the final reward function can be obtained. The reward function, R, of the UAV is shown in (15).

$$R = \beta_1 R_d + \beta_2 R_{co} + \beta_3 R_s + \beta_4 R_{fail} \tag{15}$$

where $\beta_1 + \beta_2 + \beta_3 + \beta_4 = 1$.

4. Experimental-Simulation Analysis

4.1. Simulation-Scene Settings

Based on environment modeling, a simulation experiment of multi-UAV-cooperative penetration was carried out. To simplify the training, the scene and the speed of the UAV are all set to smaller values. The number of UAVs in the cluster is set to two. In the environment, there are as many dynamic interceptors as there are UAVs. Each dynamic interceptor is responsible for the tracking of an unmanned aerial vehicle, that is, each interceptor calculates the angular velocity of the proportional guidance rotation for an unmanned aerial vehicle. The initial positions of the two interceptors are at the center of the target point, and the radius of the target area is set to 60 m. Table 1 shows the simulation parameters during training.

Table 1. Simulation Parameters.

	Parameters	Value
	Linear velocity v	20 m/s
	Angular velocity ω	-0.5 rad/s ~ 0.5 rad/s
UAV	Initial azimuth θ	$\frac{\pi}{4}$
	Initial position (x, y)	$x_A, y_A = [50 \text{ m}, 50 \text{ m}]$ $x_B, y_B = [50 \text{ m}, 500 \text{ m}]$
	Linear velocity v	22 m/s
	Initial azimuth ω	$\frac{5\pi}{4}$
Interceptor	Angular velocity θ	-0.5 rad/s ~ 0.5 rad/s
	Initial position (x, y)	Same as target
Target	Position (x, y)	$x_t, y_t = [850 \text{ m}, 850 \text{ m}]$
	Simulation step dt	1 s

During the training process, the learning rate of the actor network and the critic network are set to $\alpha_1 = 0.0001, \alpha_2 = 0.001$, the discount factor is set to $\gamma = 0.9$, and the action noise is set to 0.3.

4.2. Simulation Results and Analysis

After 10,000 trainings, the simulation results are shown in Figure 7.

Two UAVs can pass enough to bypass the dynamic interceptor and finally reach the target area at the same time, and the time to reach the target area will not exceed the simulation step (1 s).

Figures 8 and 9 shows the angular velocity change curves of UAVs and interceptors and the line-of-sight change curves of the interceptors.

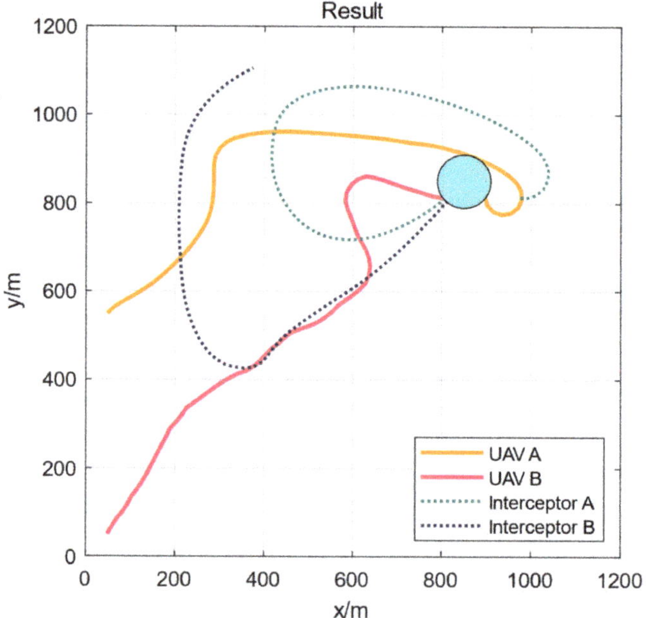

Figure 7. Experiment Result.

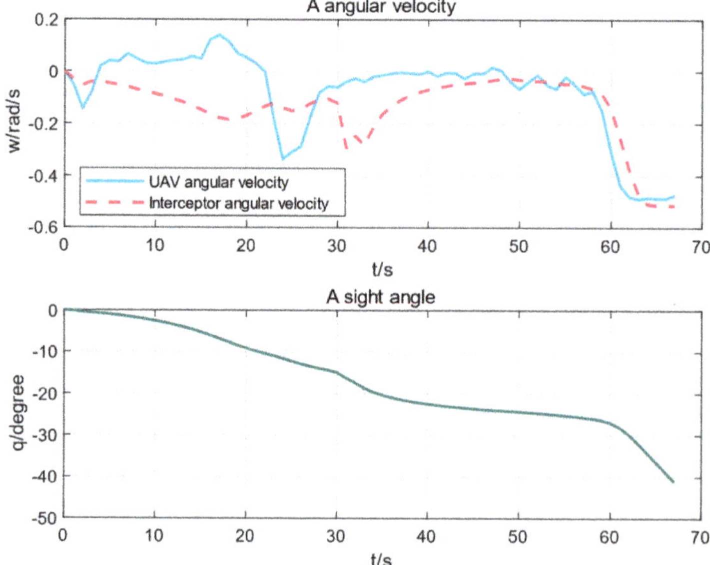

Figure 8. The parameter change curve of UAV A and Interceptor A during training.

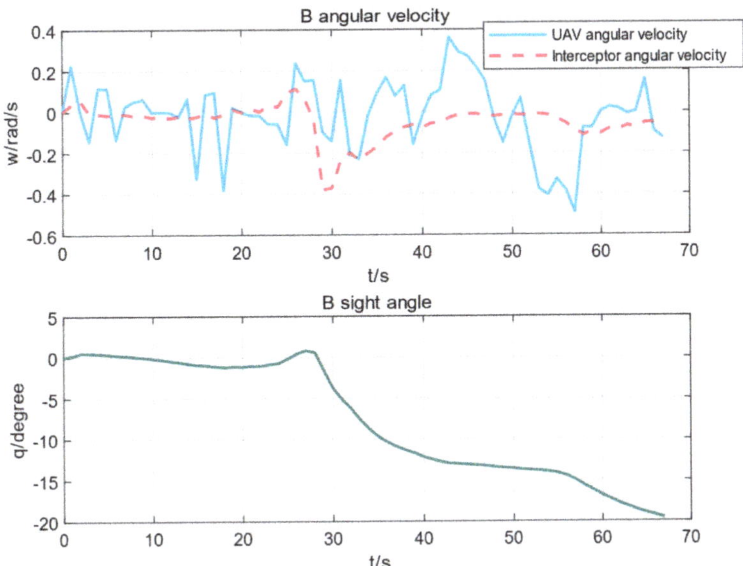

Figure 9. The parameter change curve of UAV B and Interceptor B during training.

It can be seen from the figure that the UAVs carried out a wide-angle exercise, the interceptor's line of sight to the UAVs continued to increase, and the interception failed.

Figure 10 shows the distance curve between the two UAVs and the interceptor. The black line at the bottom of the figure represents the distance captured by the interceptor.

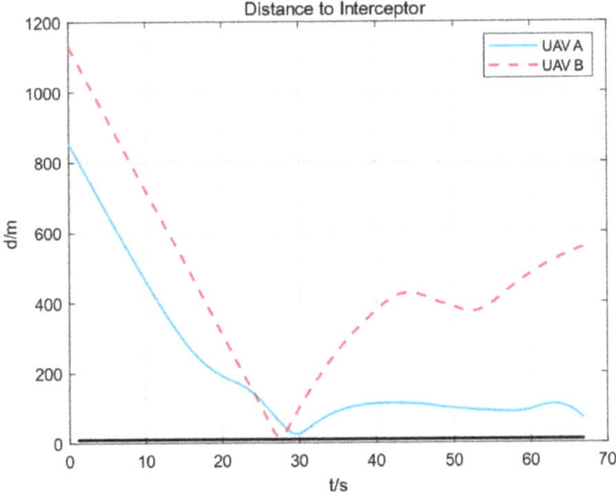

Figure 10. The distance between the UAV and the interceptor.

It can be seen from the figure that the minimum distance between the two UAVs and the interceptors is above the black line, that is, they are not captured by the interceptors.

Figure 11 is the distance between the two UAVs and the target point. In the figure, there is a certain gap between the initial distance between the two UAVs and the target area, which is about 10 m.

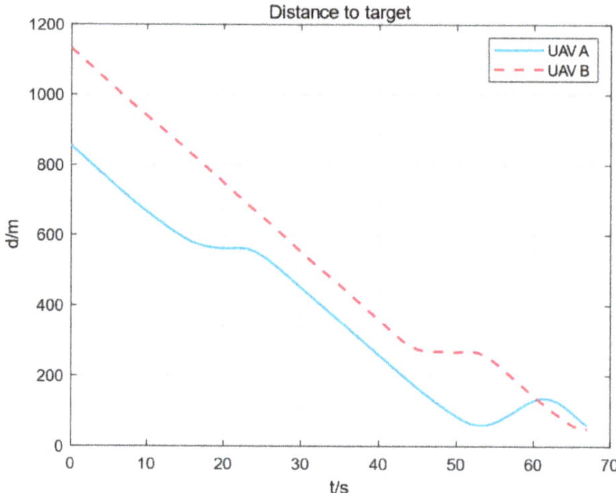

Figure 11. The distance curve between the UAV and the target.

It can be seen from the figure that the final distance difference between the two UAVs and the target point is almost zero. At the beginning of the movement, the distance between the two UAVs and the target point is about 3 m. UAV A will deliberately go around to ensure that it is moving. Finally, the target point can be reached at the same time.

The simulation experiment shows that the DDPG algorithm can complete the cooperative penetration of the dynamic-tracking interceptor through the training of the UAV cluster, which meets the high-performance requirements of the multi-UAV system and is a method with huge application potential.

5. Conclusions

Based on the deep reinforcement-learning DDPG algorithm, this paper studies the UAV-cooperative penetration. Based on the mission scenario model, the UAV's action space, state space, termination conditions, and reward functions are designed. Through the training of the UAV, the coordinated penetration of multiple aircraft was realized. The main conclusions of this paper are as follows:

The collaborative method based on the DDPG algorithm designed in this paper can achieve coordinated penetration between UAVs. After UAVs are trained, they can coordinate to evade interceptors without being intercepted by them. Compared with traditional algorithms, the UAV's penetration performance is stronger, the applicable environment is more complex, and it has great application prospects.

The reinforcement learning collaboration method in this paper realizes the time collaboration between UAVs on the premise of exchanging state information between multiple aircrafts, and the movement of UAVs finally reaches the target area at the same time from different initial locations, achieving time coordination. However, this paper doesn't consider factors such as communication delays or failures between UAVs, which causes the UAV to receive the wrong information. This problem may be solved by predicting UAVs' states and information fusion. Further research can be carried out in the follow-up work.

This paper only considers the movement of the engagement plane. In the follow-up work, the movement can be expanded to three dimensions, and multi-UAV-cooperative penetration in a 3D environment will put forward higher requirements on the algorithm.

Author Contributions: Conceptualization, J.S. and Y.L. (Yuxie Luo); methodology, Y.L. (Yuxie Luo) and K.Z.; software, Y.L. (Yuxie Luo); validation, Y.L. (Yuxie Luo), K.Z. and Y.L. (Yang Liu); formal analysis, J.S.; investigation, Y.L. (Yuxie Luo); resources, K.Z. and Y.L. (Yang Liu); data curation, J.S.; writing—original draft preparation, Y.L. (Yuxie Luo); writing—review and editing, Y.L. (Yuxie Luo);

visualization, Y.L. (Yuxie Luo); supervision, J.S.; project administration, J.S.; funding acquisition, J.S. All authors have read and agreed to the published version of the manuscript.

Funding: This research was funded by National Natural Science Foundation of China (General Program) grant number (62073020).

Institutional Review Board Statement: Not applicable.

Informed Consent Statement: Not applicable.

Data Availability Statement: The data presented in this study are available on request from the corresponding author.

Conflicts of Interest: The authors declare no conflict of interest.

References

1. Gupta, L.; Jain, R.; Vaszkun, G. Survey of important issues in UAV communication networks. *IEEE Commun. Surv. Tutor.* **2015**, *18*, 1123–1152. [CrossRef]
2. Ure, N.K.; Inalhan, G. Autonomous control of unmanned combat air vehicles: Design of a multimodal control and flight planning framework for agile maneuvering. *IEEE Control. Syst. Mag.* **2012**, *32*, 74–95.
3. Wang, K.; Song, M.; Li, M.J.S. Cooperative Multi-UAV Conflict Avoidance Planning in a Complex Urban Environment. *Sustainability* **2021**, *13*, 6807. [CrossRef]
4. Scherer, J.; Yahyanejad, S.; Hayat, S.; Yanmaz, E.; Andre, T.; Khan, A.; Vukadinovic, V.; Bettstetter, C.; Hellwagner, H.; Rinner, B. An autonomous multi-UAV system for search and rescue. In Proceedings of the First Workshop on Micro Aerial Vehicle Networks, Systems, and Applications for Civilian Use, Florence, Italy, 18 May 2015; pp. 33–38.
5. Chen, J.; Xiao, K.; You, K.; Qing, X.; Ye, F.; Sun, Q. Hierarchical Task Assignment Strategy for Heterogeneous Multi-UAV System in Large-Scale Search and Rescue Scenarios. *Int. J. Aerosp. Eng.* **2021**, *2021*, 7353697. [CrossRef]
6. Xu, Q.; Ge, J.; Yang, T. Optimal Design of Cooperative Penetration Trajectories for Multiaircraft. *Int. J. Aerosp. Eng.* **2020**, *2020*, 8490531. [CrossRef]
7. Liu, Y.; Jia, Y. Event-triggered consensus control for uncertain multi-agent systems with external disturbance. *Int. J. Syst. Sci.* **2019**, *50*, 130–140. [CrossRef]
8. Cai, Y.; Xi, Q.; Xing, X.; Gui, H.; Liu, Q. Path planning for UAV tracking target based on improved A-star algorithm. In Proceedings of the 2019 1st International Conference on Industrial Artificial Intelligence (IAI), Shenyang, China, 23–27 July 2019; pp. 1–6.
9. Chen, Y.B.; Luo, G.C.; Mei, Y.S.; Yu, J.Q.; Su, X.L. UAV path planning using artificial potential field method updated by optimal control theory. *Int. J. Syst. Sci.* **2016**, *47*, 1407–1420. [CrossRef]
10. Bounini, F.; Gingras, D.; Pollart, H.; Gruyer, D. Modified artificial potential field method for online path planning applications. In Proceedings of the 2017 IEEE Intelligent Vehicles Symposium (IV), Los Angeles, CA, USA, 11–14 June 2017; pp. 180–185.
11. Aguilar, W.G.; Morales, S.; Ruiz, H.; Abad, V. RRT* GL based optimal path planning for real-time navigation of UAVs. In *Proceedings of the International Work-Conference on Artificial Neural Networks, Cadiz, Spain, 14–16 June 2017*; Springer: Cham, Switzerland, 2017; pp. 585–595.
12. Chen, Y.; Yu, J.; Su, X.; Luo, G. Path planning for multi-UAV formation. *J. Intell. Robot. Syst.* **2015**, *77*, 229–246. [CrossRef]
13. Kothari, M.; Postlethwaite, I.; Gu, D.-W. Multi-UAV path planning in obstacle rich environments using rapidly-exploring random trees. In Proceedings of the 48h IEEE Conference on Decision and Control (CDC) Held Jointly with 2009 28th Chinese Control Conference, Shanghai, China, 15–18 December 2009; pp. 3069–3074.
14. Li, K.; Wang, J.; Lee, C.-H.; Zhou, R.; Zhao, S. Distributed cooperative guidance for multivehicle simultaneous arrival without numerical singularities. *J. Guid. Control. Dyn.* **2020**, *43*, 1365–1373. [CrossRef]
15. Ruan, W.-y.; Duan, H.-B. Multi-UAV obstacle avoidance control via multi-objective social learning pigeon-inspired optimization. *Front. Inf. Technol. Electron. Eng.* **2020**, *21*, 740–748. [CrossRef]
16. Li, Y.J. Deep reinforcement learning: An overview. *arXiv* **2017**, arXiv:1701.07274.
17. François-Lavet, V.; Henderson, P.; Islam, R.; Bellemare, M.G.; Pineau, J. An introduction to deep reinforcement learning. *arXiv* **2018**, arXiv:1811.12560.
18. Watkins, C.J.C.H. Learning from Delayed Rewards. Ph.D. Thesis, King's College, Wilkes-Barre, PA, USA, 1989.
19. Pham, H.X.; La, H.M.; Feil-Seifer, D.; Nguyen, L.V. Autonomous uav navigation using reinforcement learning. *arXiv* **2018**, arXiv:1801.05086.
20. Wang, C.; Wang, J.; Zhang, X.; Zhang, X. Autonomous navigation of UAV in large-scale unknown complex environment with deep reinforcement learning. In Proceedings of the 2017 IEEE Global Conference on Signal and Information Processing (GlobalSIP), Montreal, QC, Canada, 14–16 November 2017; pp. 858–862.
21. Wang, T.; Qin, R.; Chen, Y.; Snoussi, H.; Choi, C. A reinforcement learning approach for UAV target searching and tracking. *Multimed. Tools Appl.* **2019**, *78*, 4347–4364. [CrossRef]
22. Yang, J.; You, X.; Wu, G.; Hassan, M.M.; Almogren, A.; Guna, J. Application of reinforcement learning in UAV cluster task scheduling. *Future Gener. Comput. Syst.* **2019**, *95*, 140–148. [CrossRef]

applied sciences

MDPI

Article

Optimal Unmanned Ground Vehicle—Unmanned Aerial Vehicle Formation-Maintenance Control for Air-Ground Cooperation

Jingmin Zhang [1,2], Xiaokui Yue [1,*], Haofei Zhang [2] and Tiantian Xiao [2]

1   School of Astronautics, Northwestern Polytechnical University, Xi'an 710072, China; zjm_208suo@163.com
2   No.208 Research Institute of China Ordnance Industries, Beijing 102202, China; jeamszhf@163.com (H.Z.); tessaxiao@163.com (T.X.)
*   Correspondence: xkyue@nwpu.edu.cn

Abstract: This paper investigates the air–ground cooperative time-varying formation-tracking control problem of a heterogeneous cluster system composed of an unmanned ground vehicle (UGV) and an unmanned aerial vehicle (UAV). Initially, the structure of the UAV–UGV formation-control system is analyzed from the perspective of a cooperative combat system. Next, based on the motion relationship between the UAV–UGV in a relative coordinate system, the relative motion model between them is established, which can clearly reveal the physical meaning of the relative motion process in the UAV–UGV system. Then, under the premise that the control system of the UAG is closed-loop stable, the motion state of the UGV is modeled as an input perturbation. Finally, using a linear quadratic optimal control theory, a UAV–UGV formation-maintenance controller is designed to track the reference trajectory of the UGV based on the UAV–UGV relative motion model. The simulation results demonstrate that the proposed controller can overcome input perturbations, model-constant perturbations, and linearization biases. Moreover, it can achieve fast and stable adjustment and maintenance control of the desired UAV–UGV formation proposed by the cooperative combat mission planning system.

Keywords: UAV–UGV; cooperative engagement; optimal control; time-varying output formation; formation keeping

Citation: Zhang, J.;Yue, X.; Zhang, H. Xiao, T. Optimal Unmanned Ground Vehicle—Unmanned Aerial Vehicle Formation-Maintenance Control for Air-Ground Cooperation. *Appl. Sci.* **2022**, *12*, 3598. https://doi.org/10.3390/app12073598

Academic Editor: Yosoon Choi

Received: 6 March 2022
Accepted: 29 March 2022
Published: 1 April 2022

Publisher's Note: MDPI stays neutral with regard to jurisdictional claims in published maps and institutional affiliations.

1. Introduction

In the wake of rapid development of information technology (IT), artificial intelligence (AI), and unmanned equipment, as well as their common application in the military field, the traditional combat style, in which each platform uses its own sensors and weapon systems to detect, track, and strike targets individually, can no longer meet the needs of digital warfare. As an emerging combat style, cooperative operations can organically link various geographically dispersed sensors, command and control systems, and weapons systems into a cross-platform information network. In this manner, it is possible to connect and share battlefield information, obtain real-time situational awareness of the battlefield, and improve the integrated combat effectiveness. Cooperative combat has received increasing academic attention in recent years. However, current research on cooperative operations mainly focuses on multi-missile cooperative guidance [1–3], formation control [4–6], formation design [7–9], path planning [10–12], and multitarget assignment [13–15]. Research on air–ground cooperative operations is currently lacking.

Unmanned aerial vehicles (UAVs) and unmanned ground vehicles (UGVs) are the two most representative objects in a cluster system. Although UAVs can quickly reconnoiter a wide area, they are limited by endurance and flight altitude, which prevents them from carrying out their given tasks in special environments. For UGVs, they are able to approach targets at a close range and have a high endurance, but are slower and have a limited field of view. The advantages and disadvantages of UAVs and UGVs are summarized in Table 1. For a UAV–UGV cooperative formation, they can make full use of the advantages of spatial distribution, parallel execution of tasks, functional distribution, and fault tolerance, so that they can complement each other to compensate for the shortcomings of a single type of object and effectively improve the outcome of cooperative operations. For example, in the reconnaissance of complex terrain conditions, combat reconnaissance vehicles are often unable to detect effectively owing to obstacle blockages in the surrounding environment, and may even suffer communication losses between multiple vehicles. Introducing multiple UAVs to allow for formation control to achieve companion-flight cooperative reconnaissance can provide a large range of environmental information for the overall reconnaissance decisions of ground vehicles. Additionally, UAVs can be used as a communication relay in the case of ground communication obstacles to realize the complementary advantages and cooperation of the cluster system [16,17].

Table 1. Advantages and disadvantages of UAVs and UGVs.

	Type of UAV	Advantages	Disadvantages
UAV	Small fixed-wing UAV	Fast velocity Wide field-of-view Excellent communication	Low load capacity Low observation accuracy
	Small rotary-wing UAV	Vertical takeoff landing Good reconnaissance	Low load capacity
UGV		High load capacity Precise observation of ground targets	Small field-of-view Low velocity Weak communication

Based on this demand for engineering applications, many scholars have conducted research on the air–ground cooperative control between UAVs and UGVs. As an emerging research field, its theoretical knowledge and technical aspects involve the intersection of several disciplines and technical fields, such as embedded systems, aerodynamics, and control theory, which includes path tracking, formation clustering, automatic landing, and other research fields characterized by many research contents and difficulties [18,19]. This highlights that many problems need to be solved and explored in this area. In particular, the heterogeneity of UAVs and UGVs, for example, having different working fields, different physical models, and different technical indicators [20–22], poses new challenges to cooperative control between UAVs and UGVs [23,24].

Hence, this study investigates the air–ground cooperative time-varying formation-keeping control problem of a heterogeneous cluster system composed of a UGV and a UAV in the scenario of air–ground cooperative combat, which enables the UAV to track the motion trajectory of the UGV.

2. Related Work

Grocholsky et al. [25] proposed the use of an air–ground cooperative model to address the shortcomings of a single platform, e.g., the low accuracy of UAVs when locating ground targets and the narrow view of UGVs when observing distant obstacles. They designed a cooperative control framework and algorithm to search for and locate targets in a specified area, providing ideas for cooperative air–ground platforms to search for targets. Manyam et al. [26] considered the problem of communication constraints between two UAVs that cooperate to complete a mission. To address this issue, they proposed using UAVs to collect information and developed a branch-and-cut algorithm to solve the path-planning problem of UGVs and UAVs, offering a new idea for the cooperative path planning of UGVs and UAVs under communication constraints. In addition, they presented a cooperative reconnaissance and data-collection method for UGVs and UAVs.

To address the problem that the line of sight of a ground vehicle's sensors is limited by the geometry of the environment, Peterson et al. [27] presented a method to capture aerial images of the mission area using multi-rotors, construct a terrain map, and then navigate the ground robot to the destination. This method provides a theoretical and practical basis for the application of air–ground cooperative systems for cooperative navigation in outdoor environments.

However, the aforementioned studies on air–ground cooperation focus mostly on image transmission, map construction, and target localization, and do not consider the impact of UAV–UGV formation on the overall quality of cooperative operations. In fact, when UGVs and UAVs perform cooperative combat missions, formation adjustments are required, including initial formation generation, formation maintenance, contraction, expansion, and reconfiguration. Therefore, research on the formation control of UGVs and UAVs is of great significance, and can even affect whether a combat mission can be successfully completed.

UAVs and UGVs have completely different physical structures, and their established kinematic and dynamic models are entirely different. Moreover, UAVs perform three-dimensional movements in air, whereas UGVs perform two-dimensional movements on the ground. As a result, studying the time-varying formation-tracking control of heterogeneous cluster systems is a key problem to solve in cross-media heterogeneous collaboration, including air–ground collaboration, which has great theoretical research value and engineering significance.

Currently, common formation-control strategies include behavior-, virtual structure-, and consistency-based methods [28–30]. However, behavior-based formation methods rely on qualitative behavior rules, making it difficult to establish a quantitative model of the entire system. Consequently, such a method cannot guarantee the stability of the formation movement of an entire system. Virtual structure-based methods require centralized control by a central node, and cannot be implemented in a distributed manner. In addition, most existing methods can only achieve time-invariant formation configurations, whereas in practical applications, heterogeneous cluster systems must be able to dynamically adjust their formations in real time to cope with complex external environments and changes in tasks.

In the past, scholars have mainly used two methods to establish the relative motion models of UGVs and UAVs. One method is to directly analyze the geometric relationship of the motion of a single vehicle in a two-dimensional plane and establish a relative-motion model in a two-dimensional plane. The other is to derive the difference between the different vehicle positions in an inertial coordinate system and then obtain an expression of the relative motion state in the inertial space. This description cannot directly reflect the motion characteristics of UGVs and UAVs in a relative coordinate system, and their relative motion processes are insufficiently clear.

Based on the above research background, this study investigates the UAV–UGV formation-maintenance control problem for cooperative combat mission requirements. First, from the perspective of a UAV–UGV cooperative combat system, the architecture of the control system and the relationship between the sub-modules are analyzed. Subsequently, a relative motion model between the UGV and the UAV is established. The optimal UAV–UGV cooperative formation-maintenance control is then realized based on the proportional-integral (PI) optimal control theory. The main contributions of this study are as follows:

(1) By directly studying the motion relationship between a UGV and a UAV in a relative coordinate system, their relative equations of motion are established in three-dimensional space. It is possible to directly obtain the motion of the UGV and UAV in a relative three-dimensional coordinate system, thus clarifying the physical meaning of the relative motion between the two.

(2) The PI optimal control theory is used to design an optimal formation-maintenance controller that can overcome the constant relative-motion perturbations, as well as the

nonlinear-model linearization bias. This controller can potentially achieve fast and stable optimal control of UAV–UGV formation.

3. Relative Kinematic-Equation Building for UAV–UGV Formations

In this study, we investigate the UAV–UGV formation-maintenance control problem for cooperative combat mission requirements. Considering the actual motion state of a UGV as the input perturbation of the formation-maintenance controller, we assume that the control system of the UAV is closed-loop stable, i.e., the directions of the UAV velocity command, trajectory-inclination command, and trajectory-declination command can be stably followed. The three channels were set in the following first-order system [31]:

$$\begin{cases} \dot{V}_f = -\frac{1}{\tau_{vf}}\left(V_f - V_{fc}\right) \\ \dot{\theta}_f = -\frac{1}{\tau_{\theta f}}\left(\theta_f - \theta_{fc}\right) \\ \dot{\psi}_{vf} = -\frac{1}{\tau_{\psi vf}}\left(\psi_{vf} - \psi_{vfc}\right) \end{cases} \tag{1}$$

where τ_{vf} denotes the inertia time constant of the velocity control channel of the UAV, $\tau_{\theta f}$ denotes the inertia time constant of the track-inclination control channel of the UAV, $\tau_{\psi vf}$ denotes the inertia time constant of the track-declination control channel of the UAV, V_f is the actual velocity magnitude of the UAV, θ_f is the actual inclination angle of the UAV, ψ_{vf} is the actual declination angle of the UAV, V_{fc} is the commanded velocity magnitude of the UAV, θ_{fc} is the commanded inclination angle of the UAV, and ψ_{vfc} is the commanded declination angle of the UAV.

To study the relative motion between the UGV and UAV, the following coordinate systems are first defined:

(1) Inertial coordinate system $O_1 X_1 Y_1 Z_1$: The origin O_1 is located at a fixed point on the ground, the $O_1 X_1$ axis points to the target, the $O_1 Y_1$ axis is vertically upward, and the $O_1 Z_1$ axis forms a right-handed coordinate system with the first two axes.

(2) Relative coordinate system $O_r X_r Y_r Z_r$: The origin O_r is located at the center of mass of the UGV, the $O_r X_r$ axis points in the direction of the UGV velocity, the $O_r Y_r$ axis is in the vertical plane perpendicular to the $O_r X_r$ axis, and the $O_r Z_r$ axis forms a right-handed coordinate system with the first two axes.

(3) Vehicle-body coordinate system $O_2 X_2 Y_2 Z_2$: The origin O_2 is located at the center of mass of the UAV, the $O_2 X_2$ axis points in the direction of the UAV velocity, the $O_2 Y_2$ axis is in the vertical plane perpendicular to the $O_2 X_2$ axis, and the $O_2 Z_2$ axis forms a right-handed coordinate system with the first two axes.

In the relative coordinate system, according to Coriolis theory, the motions of the UGV and UAV are related as follows:

$$\frac{d\boldsymbol{r}}{dt} = \boldsymbol{V}_{fr} - \boldsymbol{V}_{lr} = \boldsymbol{V}_r + \boldsymbol{\omega} \times \boldsymbol{r} \tag{2}$$

where $\boldsymbol{r}$ is the relative radius vector between the UGV and UAV; $\boldsymbol{V}_{fr}$ and $\boldsymbol{V}_{lr}$ are the velocity vectors of the UAV and UGV in the relative coordinate system, respectively; $\boldsymbol{V}_r$ is the derivative of the radius vector $\boldsymbol{r}$ in the relative coordinate system; $\boldsymbol{\omega}$ is the angular velocity of rotation of the relative coordinate system in the inertial coordinate system.

To obtain the absolute velocity of the UGV and UAV in the relative coordinate system, the following transformation can be performed:

$$\begin{cases} \boldsymbol{V}_{ft} = \Phi_1^r \Phi_2^l \boldsymbol{V}_{f2} \\ \boldsymbol{V}_{lr} = \boldsymbol{V}_{l2} \end{cases} \tag{3}$$

where Φ_1^r is the conversion matrix from the inertial to the relative coordinate system; Φ_2^l is the conversion matrix from the UAV body coordinate system to the inertial coordinate system; $\boldsymbol{V}_{f2}$ is the velocity vector under the UAV body coordinate system; $\boldsymbol{V}_{l2}$ is the velocity vector under the UGV body coordinate system. The values of each of these variables are as follows:

$$\Phi_1^r = \begin{bmatrix} \cos\theta_l \cos\psi_{vl} & \sin\theta_l & -\cos\theta_l \sin\psi_{vl} \\ -\sin\theta_l \cos\psi_{vl} & \cos\theta_l & \sin\theta_l \sin\psi_{vl} \\ \sin\psi_{vl} & 0 & \cos\psi_{vl} \end{bmatrix} \tag{4}$$

$$\Phi_2^l = \begin{bmatrix} \cos\theta_f \cos\psi_{vf} & -\sin\theta_f \cos\psi_{vf} & \sin\psi_{vf} \\ \sin\theta_f & \cos\theta_f & 0 \\ -\cos\theta_f \sin\psi_{vf} & \sin\theta_f \sin\psi_{vf} & \cos\psi_{vf} \end{bmatrix} \tag{5}$$

$$V_{l2} = \begin{bmatrix} V_l \\ 0 \\ 0 \end{bmatrix} \tag{6}$$

$$V_{f2} = \begin{bmatrix} V_f \\ 0 \\ 0 \end{bmatrix} \tag{7}$$

where V_l denotes the velocity magnitude of the UGV in the inertial coordinate system; θ_l denotes the inclination angle of the UGV; ψ_{vl} denotes the deflection angle of the UGV; V_f is the velocity magnitude of the UAV in the inertial coordinate system; θ_f is the inclination angle of the UAV trajectory; ψ_{vf} is the deflection angle of the UAV trajectory. The relative velocities of the UGV and UAV in the relative coordinate system are expressed as follows:

$$V_r = \begin{bmatrix} \dot{x} \\ \dot{y} \\ \dot{z} \end{bmatrix} \tag{8}$$

where x, y, and z are the components of the position vector r of the UAV along each axis in the relative coordinate system. The angular velocity of rotation ω in the relative coordinate system with respect to the inertial space can be expressed as follows:

$$\omega = \begin{bmatrix} \dot{\psi}_{vl} \sin\theta_l \\ \dot{\psi}_{vl} \cos\theta_l \\ \dot{\theta}_l \end{bmatrix} \tag{9}$$

Then, based on Equation (2), the following equation can be obtained:

$$V_{fr} = \left(V_{fr} - V_{lr} \right) - \omega \times r \tag{10}$$

The relative motion relationship between the UAV and UGV in the three-dimensional plane can be obtained by combining Equations (2), (8), and (10).

$$\begin{cases} \dot{x} = V_f \cos\theta_f \cos\theta_l \cos\psi_e + V_f \sin\theta_f \sin\theta_l - V_l + y\dot{\theta}_l - z\dot{\psi}_{vl} \cos\theta_l \\ \dot{y} = -V_f \cos\theta_f \sin\theta_l \cos\psi_e + V_f \sin\theta_f \sin\theta_l - x\dot{\theta}_l + z\dot{\psi}_{vl} \sin\theta_l \\ \dot{z} = V_f \cos\theta_f \sin\psi_e - y\dot{\psi}_{vl} \sin\theta_l + x\dot{\psi}_{vl} \cos\theta_l \\ \psi_e = \psi_{vl} - \psi_{vf} \end{cases} \tag{11}$$

4. Optimal Control Modeling for the UAV–UGV Formation-Maintenance Controller

The design goal of the UAV–UGV formation-maintenance controller is to produce flight commands $V_{fc}, \theta_{fc}, \psi_{vf}$ for the UAV, to keep it at the desired relative distance from the UGV.

4.1. Linearization of the Relative Equations of Motion

During the UAV–UGV formation-maintenance process, assuming that θ_f, θ_l, and $\psi_e = \psi_{vl} - \psi_{vf}$ can be considered as small quantities, while considering the state of the UGV as an input quantity, Equation (11) is converted to

$$
\begin{cases}
\dot{x} = V_f + V_f \theta_f \theta_l - V_l + y\dot{\theta}_l - z\dot{\psi}_{vl} \\
\dot{y} = -V_f \theta_l + V_f \theta_f - x\dot{\theta}_l + z\dot{\psi}_{vl}\theta_l \\
\dot{z} = V_f \psi_e - y\dot{\psi}_{vl}\theta_l + x\dot{\psi}_{vl} \\
\psi_e = \psi_{vl} - \psi_{vf}
\end{cases}
\tag{12}
$$

Using the small-perturbation linearization method, Equation (12) can be linearized as

$$
\begin{cases}
\dot{x} = \dot{\theta}_l y - \dot{\psi}_{vl} z + \left(1 + \theta_{f0}\theta_l\right)V_f + V_{f0}\theta_l\theta_f - V_l \\
\dot{y} = -\dot{\theta}_l x + \dot{\psi}_{vl}\theta_l z + \left(\theta_{f0} - \theta_l\right)V_f + V_{f0}\theta_f \\
\dot{z} = \dot{\psi}_{vl} x - \dot{\psi}_{vl}\theta_l y + \left(\psi_{vl} - \psi_{vf0}\right)V_f - V_{f0}\psi_{vf}
\end{cases}
\tag{13}
$$

where v_{f0}, θ_{f0}, and ψ_{vf0} are the state equilibrium points selected during linearization. Describing Equation (13) as a state-space form (SSF) yields

$$
\begin{cases}
\dot{X} = AX + BU + \tilde{B}W \\
Y = CX
\end{cases}
\tag{14}
$$

where

$$
A = \begin{bmatrix} 0 & \dot{\theta}_L & -\dot{\psi}_{vl} \\ -\dot{\theta}_l & 0 & \theta_L\dot{\psi}_{vl} \\ \dot{\psi}_{vl} & -\theta_L\dot{\psi}_{vl} & 0 \end{bmatrix}
\tag{15}
$$

$$
B = \begin{bmatrix} 1 + \theta_{f0}\theta_l & V_{f0}\theta_l & 0 \\ \theta_{f0} - \theta_l & V_{f0} & 0 \\ \psi_{vl} - \psi_{vf0} & 0 & -V_{f0} \end{bmatrix}
\tag{16}
$$

$$
C = \begin{bmatrix} 1 & 0 & 0 \\ 0 & 1 & 0 \\ 0 & 0 & 1 \end{bmatrix}
\tag{17}
$$

$$
\tilde{B} = \begin{bmatrix} 1 \\ 0 \\ 0 \end{bmatrix}
\tag{18}
$$

where $X = \begin{bmatrix} x & y & z \end{bmatrix}^T$ denotes the state variable; $U = \begin{bmatrix} V_f & \theta_f & \psi_{vf} \end{bmatrix}^T$ denotes the control variable of the formation controller, which is the motion state of the UAV; $Y = \begin{bmatrix} x & y & z \end{bmatrix}^T$ denotes the output variable; the perturbation variable is the velocity $W = V_l$ of the UGV.

4.2. Optimal Formation-Maintenance Controller Design

According to the state-space equations of the system, the controller design problem for UAV–UGV formations is a non-zero set-point output-regulation problem with constant or slow-varying perturbations. Two steps can be taken to solve this problem: the first step is to design an optimal output regulation controller that can solve constant or slow varying perturbations. The second step is to design an optimal controller to solve the non-zero set-point problem so that the UAV–UGV formation can be stably maintained at the desired formation position.

4.2.1. PI Optimal Formation-Maintenance Controller Design

The perturbed system of type shown in Equation (14), with constant or slowly varying values, can be overcome using a PI type of control, similar to that in classical control theory. This section analyzes the problem of designing PI optimal controllers with zero values as the output regulation set points.

First, the system perturbation is transformed into a perturbation of the control input, at which point, system (14) becomes

$$\dot{X} = AX + B(U + \tilde{W}) \tag{19}$$

where $\tilde{W}$ denotes the transformed form of the original system perturbation. The following equation can be obtained from system (14):

$$\tilde{W} = B^{-1}\tilde{B}W \tag{20}$$

Next, we design the PI optimal controller of the perturbed system (19), which can be augmented as

$$\begin{cases} \dot{X}_1 = A_1 X_1 + B_1 U_1 \\ Y_1 = C_1 X_1 \end{cases} \tag{21}$$

where

$$X_1 = \begin{bmatrix} X \\ U + \tilde{W} \end{bmatrix} \tag{22}$$

$$A_1 = \begin{bmatrix} A & B \\ 0 & 0 \end{bmatrix} \tag{23}$$

$$B_1 = \begin{bmatrix} 0 \\ I \end{bmatrix} \tag{24}$$

$$C_1 = \begin{bmatrix} I & 0 \end{bmatrix} \tag{25}$$

For the augmented system (21), the quadratic performance index is given as

$$J = \int_{t_0}^{t_f} \left[X_1^T Q_1 X_1 + U_1^T R_1 U_1 \right] dt \tag{26}$$

where Q_1 is the state-regulation power-coefficient matrix of the augmented system and R_1 is the control energy power coefficient matrix of the augmented system. When the system (21) is controllable, according to optimal control theory, the minimum optimal control to achieve the quadratic performance index (26) is

$$U_1^* = -R_1^{-1}B_1^T P X_1 \tag{27}$$

Let us further analyze the quadratic performance index (26). Based on the composition of the state variable X_1, Q_1 can be decomposed as follows:

$$Q_1 = \begin{bmatrix} Q & 0 \\ 0 & R \end{bmatrix} \tag{28}$$

Then, we can have

$$X_1^T Q_1 X_1 = X^T Q X + (U + \tilde{W})^T R(U + \tilde{W}) \tag{29}$$

where Q is the state-regulation power-coefficient matrix of the original system (14) and R is its control energy power coefficient matrix.

According to system (14), we can have

$$X^T Q X = X^T C^T Q_Y C X = Y^T Q_Y Y \tag{30}$$

where Q_Y is the output regulation power-coefficient matrix of the original system.

The optimal quadratic state-regulation performance index described in Equation (26) becomes the optimal quadratic output-regulation problem.

$$J = \int_{t_0}^{t_f} \left[Y^T Q_Y Y + (U + \tilde{W})^T R (U + \tilde{W}) + U_1^T R_1 U_1 \right] dt \tag{31}$$

Assuming that the perturbation $\tilde{W}$ is a slow variable, it follows that

$$\dot{\tilde{W}} = 0 \tag{32}$$

Then, we have

$$U_1 = \mathring{U} \tag{33}$$

Thus, the quadratic optimal output performance index of the system (21) can be described as

$$J = \int_{t_0}^{t_f} \left[Y^T Q_Y Y + (\tilde{U} + \tilde{W})^T R (\tilde{U} + \tilde{W}) + \mathring{U}^T R_1 \mathring{U} \right] dt \tag{34}$$

For the optimal control quantity (27), the following equation can be obtained by combining it with (33):

$$U_1^* = \mathring{U}^* = -R_1^{-1} B_1^T P X_1 \tag{35}$$

where P is the minimum solution of the Riccati optimal control equation for achieving the performance index in Equation (26). Then, Equation (35) can be further expanded as

$$\begin{aligned}
\mathring{U}^* &= -R_1^{-1} \begin{bmatrix} 0 \\ I \end{bmatrix} \begin{bmatrix} P_{11} & P_{12} \\ P_{21} & P_{22} \end{bmatrix} \begin{bmatrix} X \\ \tilde{U}^* + \tilde{W} \end{bmatrix} \\
&= -R_1^{-1} P_{21} X - R_1^{-1} P_{22} \left(\tilde{U}^* + \tilde{W} \right)
\end{aligned} \tag{36}$$

The above equation is an indicator of the minimum optimal control amount required to achieve the zero-valued set point in Equation (26).

4.2.2. Design of Non-Zero Set-Point OPTIMAL Controllers

To maintain the output quantity $Y = \begin{bmatrix} x & y & z \end{bmatrix}^T$ at a non-zero set point, the state and control input of the system at the steady state must also be non-zero. The optimal control input based on Equation (35) should take the following form.

$$\mathring{U}^* = -R_1^{-1} B_1^T P X_1 + U_0' = -K X_1 + U_0' \tag{37}$$

where U_0' is the additional control quantity that stabilizes the non-zero set point. The output equation of the augmented system is

$$Y = C_1 X_1 \tag{38}$$

From the expanded state X_1, the augmented system is

$$X_1 = \begin{bmatrix} x \\ y \\ z \\ V_f + \tilde{W}(1) \\ \tilde{W}(2) \\ \tilde{W}(3) \end{bmatrix} \tag{39}$$

To ensure that the output of the augmented system is consistent with the output of the original system, the output matrix of the augmented system is set to

$$C_1 = \begin{bmatrix} 1 & 0 & 0 & 0 & 0 & 0 \\ 0 & 1 & 0 & 0 & 0 & 0 \\ 0 & 0 & 1 & 0 & 0 & 0 \end{bmatrix} \tag{40}$$

By substituting the control quantity (37) into the augmented system (21), we obtain

$$\dot{X}_1 = (A_1 - B_1 K)X_1 + B_1 U_0' \tag{41}$$

Because the closed-loop system (41) is asymptotically stable, the following condition should exist at steady state:

$$\lim_{t \to \infty} \dot{X}_1(t) = 0 \tag{42}$$

This leads to the following asymptotically stable form.

$$0 = (A_1 - B_1 K)X_{10} + B_1 U_0' \tag{43}$$

where X_{10} is the steady-state value of state X_1.

If all eigenvalues of matrix $(A_1 - B_1 K)$ are in the left half-complex plane, then $(A_1 - B_1 K)$ is a non-singular matrix, and the following equation can be obtained by combining with Equation (43).

$$X_{10} = -(A_1 - B_1 K)^{-1} B_1 U_0' \tag{44}$$

Then, based on Equation (38), the relationship between the non-zero set point Y_{10}^* and the steady-state value X_{10} of state X_1 is obtained as

$$Y_{10}^* = C_1 X_{10} \tag{45}$$

Finally,

$$\dot{u}^* = -R_1^{-1} B_1^{\mathrm{T}} P X_1 + \left[C_1 \left(-B_1 R_1^{-1} B_1^{\mathrm{T}} P - A_1 \right)^{-1} B_1 \right]^{-1} Y_{10}^* \tag{46}$$

can be used to achieve optimal control of the relative motion system between the UGV and the UAV, as described in Equation (14); thus, maintaining the UAV–UGV formation in the desired state.

5. Simulation Analysis

In this section, we describe the numerical simulations performed for the UAV–UGV formation-maintenance control system consisting of a UGV with a given motion state, a UAV with a first-order stability control system, and an optimal maintenance controller. The inertial time constants of the three channels of the UGV and UAV are assumed to be $\tau_{vf} = 3$ s, $\tau_{\theta f} = 1$ s, and $\tau_{\psi_{vf}} = 1$ s. The initial positions of the UAV and UGV in the inertial coordinate system are chosen as $(-5, 20, 10)$ m and $(0, 0, 0)$ m, respectively. It is assumed that the UAV–UGV formation moves for 150 s, and the simulation step size is taken to be 0.01 s. Under the action of Equation (43), the entire motion of the UAV is adjusted and maintained from the initial position to the desired position at $(-5, 20, 10)$ m. In all simulations, the required control system power coefficient matrix in Equation (34) appears as the matrices shown in Equation (47). Considering the feasible motion envelope of the

UAV, limits are set for the control amount of the formation controller and commanded motion state of the UAV, as shown in Equation (52).

$$Q_y = \begin{bmatrix} 4 & 0 & 0 \\ 0 & 6 & 0 \\ 0 & 0 & 6 \end{bmatrix}, R = \begin{bmatrix} 1 & 0 & 0 \\ 0 & 1 & 0 \\ 0 & 0 & 1 \end{bmatrix}, R_1 = \begin{bmatrix} 1 & 0 & 0 \\ 0 & 1 & 0 \\ 0 & 0 & 1 \end{bmatrix} \tag{47}$$

To verify the effectiveness of the proposed UGV–UAV formation-maintenance controller, we set two cases in which the UGV adopts different motion modes. The motion of the UGV is set as in Equations (48) to (51). The simulation results are shown in Figures 1 and 2.

case 1:

$$V_l(t) = \begin{cases} 2 + \sin(0.1t) \text{ m/s}, 0 \leq t < 50 \text{ s} \\ 2 \text{ m/s}, 50 \leq t < 100 \text{ s} \\ 2 + \sin(0.1t) \text{ m/s}, 100 \leq t < 150 \text{ s}. \end{cases} \tag{48}$$

$$\begin{cases} \theta_l(t) = 0 \text{ rad} \\ \psi_{vl}(t) = 0.5\cos(0.1t + 0.7728) \text{ rad} \end{cases} \tag{49}$$

case 2:

$$V_l(t) = \begin{cases} 1.5 \text{ m/s}, 0 \leq t < 50 \text{ s} \\ 2 + \sin(0.1t) \text{ m/s}, 50 \leq t < 100 \text{ s} \\ 1.5 \text{ m/s}, 100 \leq t < 150 \text{ s} \end{cases} \tag{50}$$

$$\begin{cases} \theta_l(t) = 0 \text{ rad} \\ \psi_{vl}(t) = 0.2\sin(0.15t + 0.5523) \text{ rad} \end{cases} \tag{51}$$

$$\begin{cases} 1 \text{ m/s} \leqslant V_{fc} \leqslant 3 \text{ m/s} \\ -15° \leqslant \theta_{fc} \leqslant 25° \\ -80° \leqslant \psi_{vfc} \leqslant 80° \end{cases} \tag{52}$$

It can be seen from Figures 1 and 2 that under the control command in Equation (46), the UGV can guide the UAV to fly at a fixed altitude in accordance with the specified motion, which results in the maintenance of a fixed formation in the three-dimensional plane. In addition, the velocity and declination angle commands of the UVA vary slightly and reasonably without the jitter vibration phenomenon, which satisfies the constraints of the laws of physics. Moreover, the motion of the UGV perturbs the UAV–UGV formation-maintenance control. As the UGV changes its motion form at 50 and 100 s, the motion trajectory of the UAV fluctuates at the corresponding moments. Nevertheless, it can be quickly stabilized at the desired motion state using a formation-maintenance controller. In case 1, the maximum values of the motion state errors of the UAV in the X_r, Y_r, and Z_r directions are 1.2 m , 0 m, and 1.8 m, respectively. In Case 2, the maximum values of the motion state errors of the UAV in the X_r, Y_r, and Z_r directions are 0.6 m, 0 m, and 0.6 m, respectively.

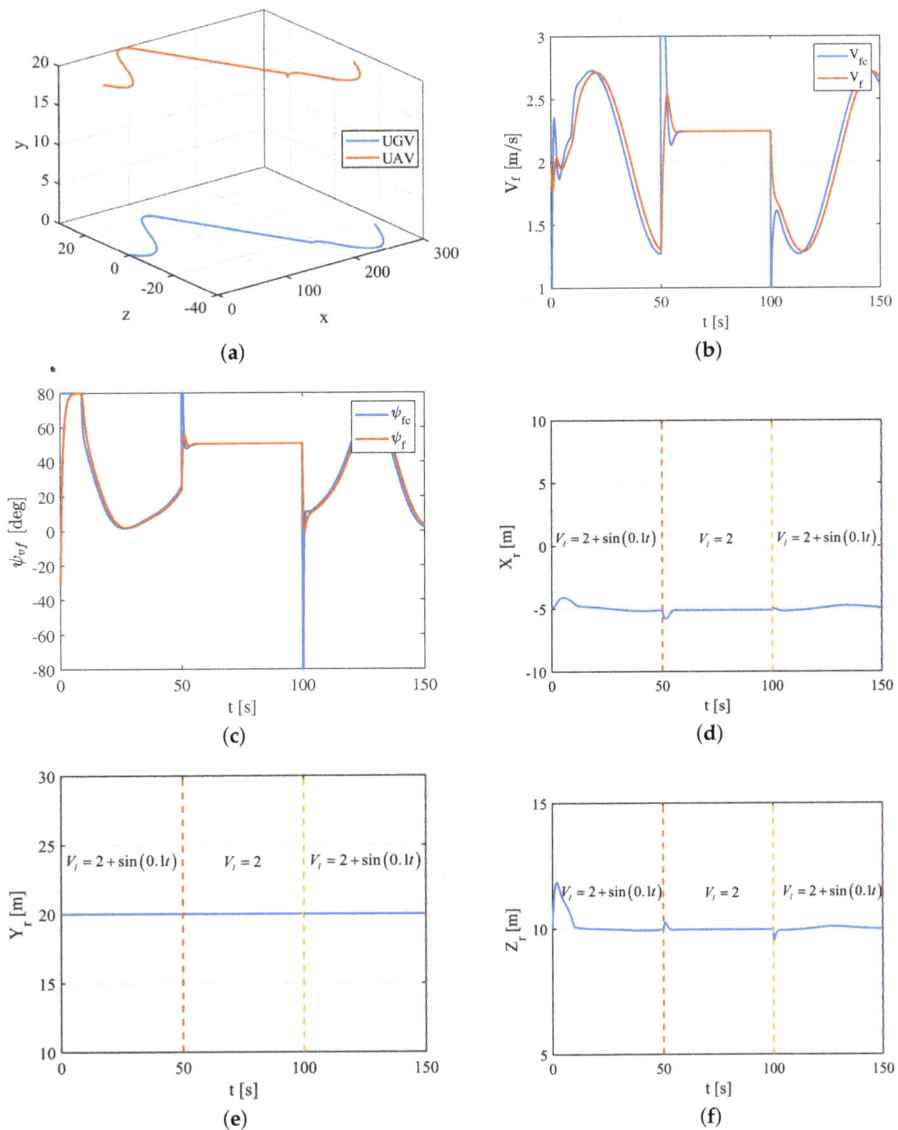

Figure 1. Simulation results of case 1. (**a**) Three-dimensional motion trajectory of UGV and UAV; (**b**) velocity of UAV; (**c**) deflection angle of UAV; (**d**) motion trajectory in X_r-axis direction; (**e**) motion trajectory in Y_r-axis direction; (**f**) motion trajectory in Z_r-axis direction.

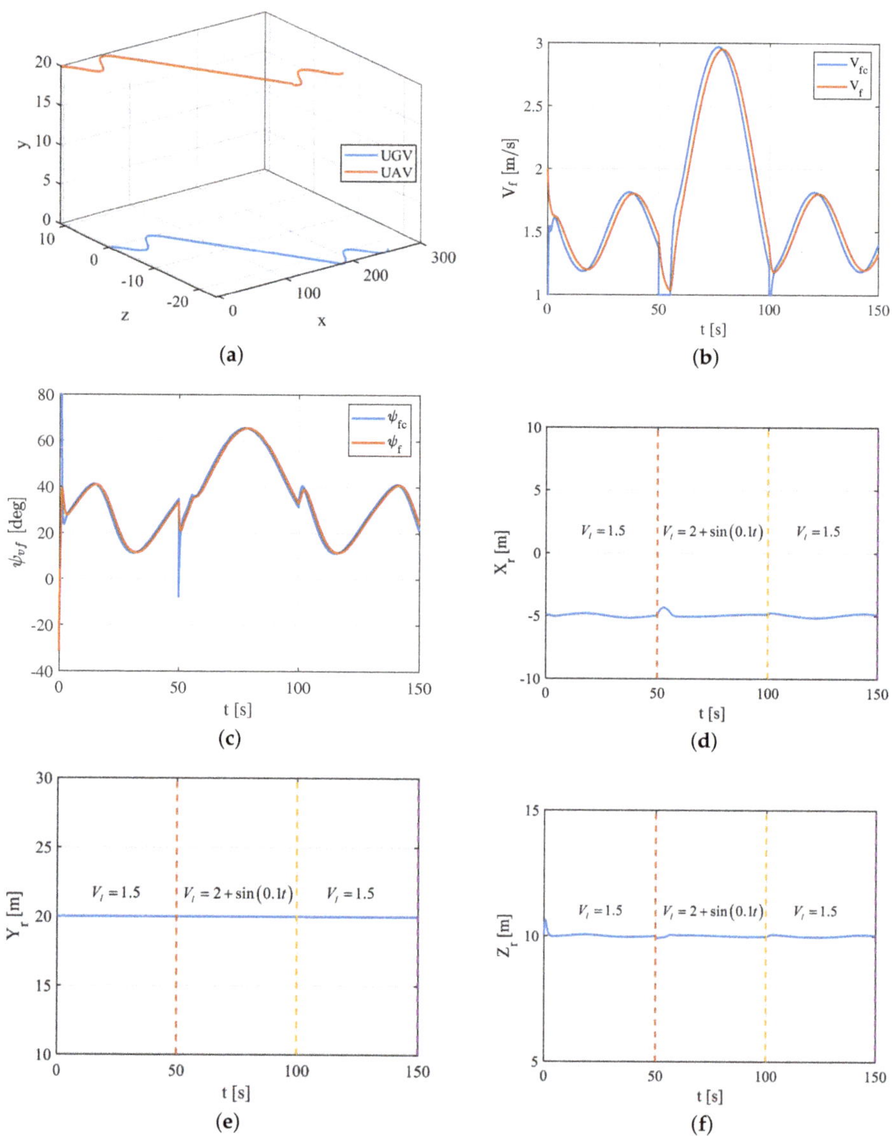

Figure 2. Simulation results of case 2. (**a**) Three-dimensional motion trajectory of UGV and UAV; (**b**) velocity of UAV; (**c**) deflection angle of UAV; (**d**) motion trajectory in X_r-axis direction; (**e**) motion trajectory in Y_r-axis direction; (**f**) motion trajectory in Z_r-axis direction.

6. Conclusions

This study analyzed the structure of the UAV–UGV formation motion control system and the relationship between the subsystems. A UAV–UGV formation–maintenance controller was designed based on the optimal control theory. In addition, simulations were performed for the UAV–UGV formation-maintenance control system, including a UGV with a given motion state, a UAV with a first-order stability-control system, and an optimal maintenance controller. The following conclusions were drawn.

(1) The physical significance of the UAV–UGV relative-motion model, based directly on the UAV motion relationship in the relative coordinate system, was clear.

(2) The optimal UAV–UGV formation-maintenance controller designed in this study had quadratic optimal properties for the UAV–UGV relative-motion state, as well as the formation-control energy. The controller could overcome the constant perturbation of the UAV–UGV relative motion caused by the velocity of the UGV. The optimal UAV–UGV formation-maintenance controller could overcome the given motion state of the UGV as an input perturbation, while the UGV performed a prolonged motion.

(3) Within the flight envelope of the UAV–UGV formation, the optimal UAV–UGV formation-maintenance controller was able to overcome the errors introduced by the linearization of a nonlinear model.

Author Contributions: The contributions of the authors are the following; conceptualization: all authors; methodology: H.Z. and X.Y.; software: T.X.; writing: J.Z.; investigation: T.X. All authors have read and agreed to the published version of the manuscript.

Funding: The authors appreciate the financial support from the National Natural Science Foundation of China (NSFC) (62173274), Natural Science Foundation of Shaanxi Province (2020JQ-219), and Aviation Fund (20200001053005).

Institutional Review Board Statement: Not applicable.

Informed Consent Statement: Not applicable.

Data Availability Statement: All data used during the study appear in the submitted article.

Conflicts of Interest: All of the authors declare that they have no known competing financial interests or personal relationships that could appear to influence the work reported in this paper.

References

1. Song, J.; Song, S.; Xu, S. Three-dimensional cooperative guidance law for multiple missiles with finite-time convergence. *Aerosp. Sci. Technol.* **2017**, *67*, 193–205. [CrossRef]
2. Liu, S.; Yan, B.; Liu, R.; Dai, P.; Yan, J.; Xin, G. Cooperative guidance law for intercepting a hypersonic target with impact angle constraint. *Aeronaut. J.* **2022**, 1–19.
3. Wang, C.; Ding, X.; Wang, J.; Shan, J. A robust three-dimensional cooperative guidance law against maneuvering target. *J. Frankl. Inst.* **2020**, *357*, 5735–5752. [CrossRef]
4. Chen, C.P.; Yu, D.; Liu, L. Automatic leader-follower persistent formation control for autonomous surface vehicles. *IEEE Access* **2018**, *7*, 12146–12155. [CrossRef]
5. Lu, Y.; Zhang, G.; Sun, Z.; Zhang, W. Adaptive cooperative formation control of autonomous surface vessels with uncertain dynamics and external disturbances. *Ocean Eng.* **2018**, *167*, 36–44. [CrossRef]
6. Kuriki, Y.; Namerikawa, T. Consensus-based cooperative formation control with collision avoidance for a multi-UAV system. In Proceedings of the 2014 American Control Conference, Linz, Austria, 15–17 July 2014; pp. 2077–2082.
7. Liu, S.; Huang, F.; Yan, B.; Zhang, T.; Liu, R.; Liu, W. Optimal Design of Multimissile Formation Based on an Adaptive SA-PSO Algorithm. *Aerospace* **2022**, *9*, 21. [CrossRef]
8. Lin, X.; Shi, X.; Li, S. Optimal cooperative control for formation flying spacecraft with collision avoidance. *Sci. Prog.* **2020**, *103*. [CrossRef]
9. Ousingsawat, J.; Campbell, M.E. Optimal cooperative reconnaissance using multiple vehicles. *J. Guid. Control Dyn.* **2007**, *30*, 122–132. [CrossRef]
10. Bellingham, J.S.; Tillerson, M.; Alighanbari, M.; How, J.P. Cooperative path planning for multiple UAVs in dynamic and uncertain environments. In Proceedings of the 41st IEEE Conference on Decision and Control, Las Vegas, NV, USA, 10–13 December 2002; Volume 3, pp. 2816–2822.
11. Yu, H.; Meier, K.; Argyle, M.; Beard, R.W. Cooperative path planning for target tracking in urban environments using unmanned air and ground vehicles. *IEEE/ASME Trans. Mechatron.* **2014**, *20*, 541–552.
12. Gang, L.; Song-Yang, L.; Dong-Feng, T.; Zhi-Chao, Z. Research status and progress on anti-ship missile path planning. *Acta Autom. Sin.* **2013**, *39*, 347–359.
13. Liu, S.; Liu, W.; Huang, F.; Yin, Y.; Yan, B.; Zhang, T. Multitarget allocation strategy based on adaptive SA-PSO algorithm. *Aeronaut. J.* **2022**, 1–13.
14. Guo, J.; Hu, G.; Guo, Z.; Zhou, M. Evaluation Model, Intelligent Assignment and Cooperative Interception in Multi-missile and Multi-target Engagement. *IEEE Trans. Aerosp. Electron. Syst.* **2022**.

15. Poore, A.P.; Rijavec, N. A Lagrangian relaxation algorithm for multidimensional assignment problems arising from multitarget tracking. *SIAM J. Optim.* **1993**, *3*, 544–563. [CrossRef]

16. Lakas, A.; Belkhouche, B.; Benkraouda, O.; Shuaib, A.; Alasmawi, H.J. A framework for a cooperative UAV–UGV system for path discovery and planning. In Proceedings of the 2018 International Conference on Innovations in Information Technology (IIT), Al Ain, United Arab Emirates, 18–19 November 2018; pp. 42–46.

17. Arbanas, B.; Ivanovic, A.; Car, M.; Orsag, M.; Petrovic, T.; Bogdan, S. Decentralized planning and control for UAV–UGV cooperative teams. *Auton. Robot.* **2018**, *42*, 1601–1618. [CrossRef]

18. Yu, K.; Budhiraja, A.K.; Buebel, S.; Tokekar, P. Algorithms and experiments on routing of unmanned aerial vehicles with mobile recharging stations. *J. Field Robot.* **2019**, *36*, 602–616. [CrossRef]

19. Harik, E.H.C.; Guérin, F.; Guinand, F.; Brethé, J.F.; Pelvillain, H. UAV–UGV cooperation for objects transportation in an industrial area. In Proceedings of the 2015 IEEE International Conference on Industrial Technology (ICIT), Seville, Spain, 17–19 March 2015; pp. 547–552.

20. Rodriguez-Ramos, A.; Sampedro, C.; Bavle, H.; De La Puente, P.; Campoy, P. A deep reinforcement learning strategy for UAV autonomous landing on a moving platform. *J. Intell. Robot. Syst.* **2019**, *93*, 351–366. [CrossRef]

21. Palafox, P.R.; Garzón, M.; Valente, J.; Roldán, J.J.; Barrientos, A. Robust visual-aided autonomous takeoff, tracking, and landing of a small UAV on a moving landing platform for life-long operation. *Appl. Sci.* **2019**, *9*, 2661. [CrossRef]

22. Ghamry, K.A.; Kamel, M.A.; Zhang, Y. Cooperative forest monitoring and fire detection using a team of UAVs-UGVs. In Proceedings of the 2016 International Conference on Unmanned Aircraft Systems (ICUAS), Arlington, VA, USA, 7–10 June 2016; pp. 1206–1211.

23. Kamel, M.A.; Ghamry, K.A.; Zhang, Y. Real-time fault-tolerant cooperative control of multiple UAVs-UGVs in the presence of actuator faults. *J. Intell. Robot. Syst.* **2017**, *88*, 469–480. [CrossRef]

24. Khaleghi, A.M.; Xu, D.; Wang, Z.; Li, M.; Lobos, A.; Liu, J.; Son, Y.J. A DDDAMS-based planning and control framework for surveillance and crowd control via UAVs and UGVs. *Expert Syst. Appl.* **2013**, *40*, 7168–7183. [CrossRef]

25. Grocholsky, B.; Keller, J.; Kumar, V.; Pappas, G. Cooperative air and ground surveillance. *IEEE Robot. Autom. Mag.* **2006**, *13*, 16–25. [CrossRef]

26. Manyam, S.G.; Casbeer, D.W.; Sundar, K. Path planning for cooperative routing of air-ground vehicles. In Proceedings of the 2016 American Control Conference (ACC), Boston, MA, USA, 6–8 July 2016; pp. 4630–4635.

27. Peterson, J.; Chaudhry, H.; Abdelatty, K.; Bird, J.; Kochersberger, K. Online aerial terrain mapping for ground robot navigation. *Sensors* **2018**, *18*, 630. [CrossRef] [PubMed]

28. Beard, R.W.; Lawton, J.; Hadaegh, F.Y. A coordination architecture for spacecraft formation control. *IEEE Trans. Control Syst. Technol.* **2001**, *9*, 777–790. [CrossRef]

29. Takahashi, H.; Nishi, H.; Ohnishi, K. Autonomous decentralized control for formation of multiple mobile robots considering ability of robot. *IEEE Trans. Ind. Electron.* **2004**, *51*, 1272–1279. [CrossRef]

30. Oh, K.K.; Park, M.C.; Ahn, H.S. A survey of multi-agent formation control. *Automatica* **2015**, *53*, 424–440. [CrossRef]

31. Cui, N.; Wei, C.; Guo, J.; Zhao, B. Research on missile formation control system. In Proceedings of the 2009 International Conference on Mechatronics and Automation, Changchun, China, 9–12 August 2009; pp. 4197–4202.

applied sciences

Article

Optimal Cruise Characteristic Analysis and Parameter Optimization Method for Air-Breathing Hypersonic Vehicle

Hesong Li [1], Yunfan Zhou [1], Yi Wang [1,*], Sha Du [2] and Shangcheng Xu [1]

[1] College of Aerospace Science, National University of Defense Technology, Changsha 410073, China; lihesong@nudt.edu.cn (H.L.); zhouyunfan@nudt.edu.cn (Y.Z.); shangchengxu@nudt.edu.cn (S.X.)
[2] Shanghai Electromechanical Engineering Research Institute, Shanghai 201109, China; dusha1983@hotmail.com
* Correspondence: wangyi@nudt.edu.cn; Tel.: +86-130-6023-3707

Abstract: There is an optimal cruise point with the lowest fuel consumption when a hypersonic vehicle performs steady-state cruise. The optimal cruise point is composed of the optimal cruise altitude and the optimal cruise Mach number, and its position is closely related to the aircraft parameters. This article aims to explore the relationship between the optimal cruise point and relevant aircraft parameters and establish a model to describe it, then an aircraft parameter optimization method of adjusting the optimal cruise point to the target position is explored with validation by numerical simulation. Firstly, a parameterized model of a hypersonic vehicle is obtained as a basis, then the optimal cruise point is obtained by the optimization method, and the influence of a single aircraft parameter on the optimal point is investigated. In order to model the relationship between the aircraft parameters and the optimal cruise point, a neural network is employed. Finally, the model is used to optimize the aircraft parameters under multiple constraints. The results show that, after aircraft parameters optimization, the optimal cruise point is located at the predetermined position and the fuel consumption is lower, which provides a new perspective for the design of aircraft.

Keywords: hypersonic vehicle; steady-state cruise; aircraft parameter; neural network

Citation: Li, H.; Zhou, Y.; Wang, Y.; Du, S.; Xu, S. Optimal Cruise Characteristic Analysis and Parameter Optimization Method for Air-Breathing Hypersonic Vehicle. *Appl. Sci.* **2021**, *11*, 9565. https://doi.org/10.3390/app11209565

Academic Editors: Giovanni Bernardini, Wei Huang and Dong Zhang

Received: 14 September 2021
Accepted: 12 October 2021
Published: 14 October 2021

Publisher's Note: MDPI stays neutral with regard to jurisdictional claims in published maps and institutional affiliations.

1. Introduction

An air-breathing hypersonic vehicle generally refers to an aircraft powered by air-breathing engines and flying at a Mach number above 5 [1]. It has a series of advantages, such as high altitude, fast speed, and strong penetrability, which has a far-reaching impact on the future development of the aerospace field [2,3]. Therefore, the research of hypersonic technology has received extensive attention from researchers all over the world.

In the cruise phase of a hypersonic vehicle, steady-state cruise refers to the cruise mode whose altitude and speed remain constant [4]. Steady-state cruise is simple and direct and has high stability for a hypersonic vehicle [5]. In order to save fuel and increase the range of the aircraft, the fuel consumption averaged by the range is an important indicator that many studies of trajectory optimization focus on. Research has shown that, when the aircraft performs steady-state cruise at different altitudes or Mach numbers, the fuel consumption is different [6–9]. Under a determined aircraft model, there is an optimal steady-state cruise point composed of the optimal cruise Mach number and the optimal cruise altitude. When the aircraft performs steady-state cruise at the Mach number and altitude of the optimal point, the fuel consumption averaged by the range is lowest [10]. To obtain the optimal cruise point, a great deal of research is carried out.

Based on a parametric model of HL-20, a hypersonic vehicle widely used in trajectory optimization research, Liu et al. [11] developed a two-level optimization algorithm combining particle swarm optimization and sequential quadratic programming to solve the optimal steady-state cruise point; Gao et al. [12] employed the fmincon function in MATLAB to determine the position of the optimal cruise point. In [13], it was pointed out that a higher altitude was beneficial to reduce drag, but, at the same time, it would reduce the specific impulse. Therefore, the optimal cruise point was a combined result of these two aspects.

Most of this research is based on the HL-20 aircraft model. However, the parametric model of HL-20 is an ideal model whose specific impulse is independent from the equivalence ratio, while it has been found that the specific impulse should be negatively correlated with the equivalent ratio in the ramjet engine [14]. As a hypersonic vehicle enters the stage of engineering practice, to better develop the hypersonic vehicle, it is significant to clarify the cruise characteristics based on a more practical aircraft model.

In addition, for different aircraft models, the positions of the optimal cruise points are also different, which reveals the position of the optimal cruise point is closely related to the aircraft parameters. Even though some methods have been developed to solve the optimal cruise point, there is little research about how the aircraft parameters influence the optimal cruise point. In the aircraft design process, when the aircraft parameters are determined, there is a corresponding optimal steady-state cruise point. However, if the subjectively desired cruise Mach number and altitude deviate from the objective Mach number and altitude of the optimal cruise point, the fuel consumption cannot be reduced fully. Then, the optimal cruise point can be taken into account in the aircraft design to guide the optimization of the aircraft parameters. The optimal cruise point can be changed after aircraft parameters optimization and then coincide with the subjective desired cruise point, which is of great benefit to save fuel. Therefore, exploring the influence law of aircraft parameters on the optimal cruise point and carrying out the research on aircraft parameter optimization is of great significance to provide a new perspective for aircraft design.

Regarding the analysis and optimization of aircraft parameters, a great deal of research has been carried out. In the early days, researchers focused on analyzing a single aircraft parameter to investigate its impact on the range, lift-to-drag ratio, and other aspects of performance [15]. The control variable method was widely adopted, especially in the research about range. To study the relationship between the range of a hypersonic vehicle and the aircraft parameters, the Bruguet formula was employed [16,17], and some methods, such as the cell mapping method in [18], were applied as well. Moreover, the aerodynamic layout of the hypersonic vehicle was analyzed to clarify the influence on the lift-drag ratio and other indicators [19,20].

However, due to the strong interaction between the aircraft parameters, the hypersonic vehicle is a complex system. The analysis and optimization of the aircraft parameters cannot focus only on a single factor. The coupling relationship between various parameters also needs to be taken into consideration. Therefore, in recent years, surrogate model technology has been widely employed in the analysis of aircraft parameters. The relationship between the aircraft parameters and performance indicators can be described by a surrogate model, and then parameter optimization can be carried out based on the surrogate model [21]. Currently, to establish a surrogate model, polynomial regression, the Kriging model, radial basis function, and a neural network can be employed [22]. In addition, many statistical methods have been introduced into the research based on the surrogate model. In the analysis of the aircraft parameters, the sensitivity analysis method [23] and correlation analysis method were widely used to dig out important parameters whose influence was most obvious [24,25]. The orthogonal test and some novel methods, such as the technology identification, evaluation and selection method in [26] and hybrid algorithm in [27], were employed to explore an optimal design space and parameter estimation in aircraft parameters analysis as well.

Based on the surrogate model, optimization methods have been widely employed in the parameter optimization research [28]. There are various parameter optimization methods, which can be mainly divided into gradient-based methods and the intelligent method. With the development of computing science, the intelligent method increasingly plays an important role. A genetic algorithm (GA), particle swarm optimization (PSO), differential evolution (DE) algorithm, and so on have been widely developed [29]. For an optimization problem with high dimensions, the PSO algorithm is demonstrated to be suitable and is also easy to program, so a great deal of research has adopted it [30,31].

In this paper, in order to describe the influence of aircraft parameters on the optimal cruise point, a sensitivity analysis and surrogate model technology with the optimization method is employed. Firstly, based on a parametric hypersonic vehicle model, the position of the optimal cruise point is quickly solved by the PSO algorithm. On this basis, the influence law of a single aircraft parameter on the optimal cruise point is explored, and a model about the relationship between the aircraft parameters and optimal cruise point is established by neural network. Based on the model, the aircraft parameters are optimized under various constraints to adjust the optimal cruise point to the given position. The aircraft parameters optimization method proposed reveals how the aircraft parameters should be adjusted, which is able to provide guidance in the design of a hypersonic vehicle.

2. Models

2.1. Dynamic Equations

For simplicity, it is considered that Earth is a homogeneous sphere with a radius of 6378 km. The Coriolis inertial force is ignored, and the gravitational acceleration is regarded as a constant at 9.8 m/s^2. The force analysis of aircraft is shown in Figure 1, where T is the thrust vector generated by the aircraft engine, R is the aerodynamic force vector that contains lift and drag, and G is the gravity vector.

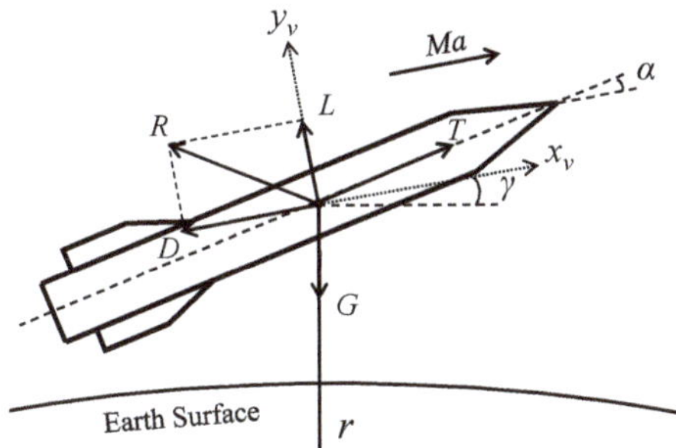

Figure 1. Force analysis diagram of aircraft.

The dynamic equation of the aircraft mass center in a form of vectors is as follows:

$$m\frac{d^2\vec{r}}{dt^2} = \vec{T} + \vec{R} + \vec{G} \tag{1}$$

In the velocity coordinate system, Equation (1) can be decomposed into

$$\begin{cases} m\frac{dv}{dt} = T\cos\alpha - D - G\sin\gamma \\ m(v \cdot \frac{d\gamma}{dt} - \frac{v^2\cos\gamma}{r}) = T\sin\alpha + L - G\cos\gamma \end{cases} \tag{2}$$

where v denotes the flight velocity, γ is the elevation angle of trajectory, α is the angle of attack, and m is the mass of the aircraft. Due to

$$\begin{cases} h = r - R_e \\ v = Ma \cdot c \end{cases} \tag{3}$$

where h denotes the flying height (km) and Ma is Mach number, then the dynamic equation of the cruise phase is [6]:

$$\begin{cases} \frac{dh}{dt} = Ma \cdot c \cdot \sin \gamma \\ \frac{dMa}{dt} = \frac{T \cos \alpha - D - mg \sin \gamma}{m \cdot c} \\ \frac{d\gamma}{dt} = \frac{T \sin \alpha + L}{m Ma \cdot c} + \cos \gamma \left(\frac{Ma \cdot c}{R_e + h} - \frac{g}{Ma \cdot c} \right) \end{cases} \tag{4}$$

Based on the dynamic equation, the trajectory can be solved by numerical method. Before that, the magnitude of lift, drag, and thrust need to be determined.

2.2. Aerodynamic Force Calculation

The magnitude of lift and drag is calculated by lift coefficient and drag coefficient, which are denoted by C_L and C_D, respectively, as follows:

$$\begin{aligned} L &= C_L \cdot \tfrac{1}{2}\rho(Ma \cdot c)^2 \cdot S \\ D &= C_D \cdot \tfrac{1}{2}\rho(Ma \cdot c)^2 \cdot S \end{aligned} \tag{5}$$

where ρ is the atmospheric density and S is the reference area of the aircraft. The lift coefficient and drag coefficient under different angles of attack, rudder angles, and incoming Mach numbers are obtained by CFD simulation. Then, in the process of trimming angle of attack, the rudder angle is adjusted to make the pitching moment equal to 0, and the lift coefficient and drag coefficient at this moment are regarded as the data under this angle of attack in steady state [32]. Subsequently, the lift coefficient and drag coefficient are parameterized by polynomial fitting. The fitting polynomials are as follows, and the values of coefficients are given in Tables 1 and 2.

$$\begin{aligned} C_L &= \sum_{i=0,j=0}^{i+j\leq 3} A_{ij} \cdot Ma^i \cdot \alpha^j \\ C_D &= \sum_{i=0,j=0}^{i\leq 3, i+j\leq 4} B_{ij} \cdot Ma^i \cdot \alpha^j \end{aligned} \tag{6}$$

Table 1. Fitting coefficients of C_L.

j \ i	0	1	2	3
0	−0.3103	0.2198	−0.0491	0.0032
1	0.1002	0.00015	−0.00069	
2	0.0032	−0.00024		
3	0.0000456			

Table 2. Fitting coefficients of C_D.

j \ i	0	1	2	3
0	0.5957	−0.1855	0.0282	−0.0015
1	−0.0470	0.0314	−0.0066	0.00044
2	−0.00086	0.00086	−0.000064	
3	0.00013	−0.000023		
4	0.00000255			

Figure 2 shows the comparison between the original data and the fitted results. It can be seen that, with the increase of angle of attack, the lift coefficient and drag coefficient increase, and, with the increase of Mach number, both of them decrease slowly.

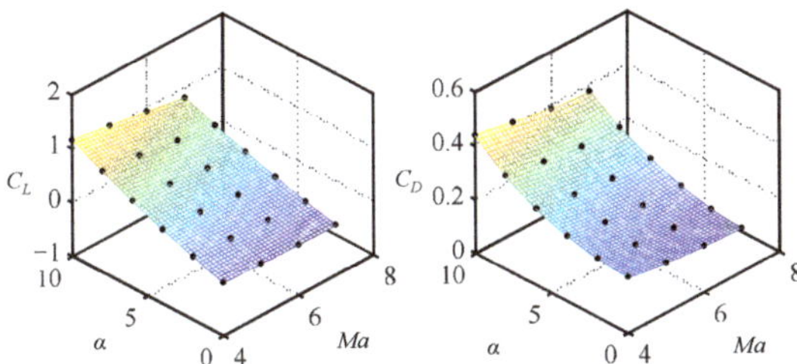

Figure 2. The change of C_L and C_D with angle of attack and Mach number.

The atmospheric parameters change a lot at different altitudes, which has a significant impact on the calculation of aerodynamic forces. In this paper, the American Standard Atmospheric Parameters Model is employed. Firstly, a gravity potential height is defined [33]:

$$H = \frac{h}{1 + h/R_e} \tag{7}$$

where R_e denotes the radius of Earth. When the altitude is within 20 to 32 km, the atmospheric density and temperature are calculated by the following formula:

$$W = 1 + \frac{H - 124.9021}{221.552}$$
$$\frac{\rho}{\rho_0} = 2.5158 \times 10^{-2} W^{-34.1629} \tag{8}$$
$$T = 221.552W \ (K)$$

where $\rho_0 = 1.225 \ \text{kg/m}^3$.

The sound velocity calculation formula is:

$$c = \sqrt{1.4 \times 287 \times T} \tag{9}$$

According to the formulas above, the lift and drag of aircraft can be calculated.

2.3. Thrust Calculation

The thrust of a hypersonic vehicle is generated by the ramjet. Air flow is captured by the inlet, then mixed with fuel, and burns to generate thrust. The air mass flow rate captured by the engine inlet, denoted by $\dot{m}_{inlet}$, is evaluated:

$$\dot{m}_{inlet} = \varphi_{inlet} \cdot \rho \cdot Ma \cdot c \cdot A \tag{10}$$

where A denotes the inlet area and φ_{inlet} is the inlet flow capture coefficient, which reflects the ability to capture air flow under different Mach numbers and angles of attack.

The calculation of φ_{inlet} is shown in Equation (11), and the values of coefficients are given in Table 3.

$$\varphi_{inlet} = K_1 \cdot \alpha^2 + K_2 \cdot Ma^2 + K_3 \cdot \alpha + K_4 \cdot Ma + K_5 \cdot Ma \cdot \alpha + K_6 \tag{11}$$

Table 3. Fitting coefficients of φ_{inlet}.

K_1	K_2	K_3	K_4	K_5	K_6
0.0001135	-0.002	-0.01232	0.033	0.00553	0.0815

Since the kind of fuel in the scramjet is fixed, there is a relationship between the fuel mass flow rate and the air mass flow rate in complete combustion, which is displayed in

Equation (12), where k is a proportion coefficient with a value at 0.07 and $\dot{m}_{fst}$ denotes the fuel mass flow rate required for complete combustion.

$$\dot{m}_{fst} = k \cdot \dot{m}_{inlet} \tag{12}$$

The ratio of current fuel mass flow rate to the fuel mass flow rate required for complete combustion is called the equivalence ratio, which is denoted by Φ [34]:

$$\Phi = \frac{\dot{m}_{fuel}}{\dot{m}_{fst}} \tag{13}$$

When $\Phi = 1$, it indicates that current amount of air flow can be completely consumed; when $\Phi < 1$, it indicates that the current combustion in the engine is in an oxygen enriched state. However, when $\Phi < 0.45$, it leads to insufficient fuel, and the engine cannot work normally. Therefore, the range of Φ is between 0.45 and 1.

Combining Equations (12) and (13), the fuel mass flow rate is:

$$\dot{m}_{fuel} = \Phi \cdot k \cdot \dot{m}_{inlet} \tag{14}$$

Then, the thrust is:

$$T = \dot{m}_{fuel} \cdot I_{sp} = \Phi \cdot k \cdot \dot{m}_{inlet} \cdot I_{sp} \tag{15}$$

where I_{sp} denotes the specific impulse, a parameter related to equivalence ratio and Mach number:

$$I_{sp} = f(\Phi) \cdot g(Ma) \tag{16}$$

$f(\Phi)$ and $g(Ma)$ are shown in Equation (17), and the values of coefficients are given in Table 4.

$$\begin{aligned} f(\Phi) &= \left(1 + \frac{1-\Phi}{2}\right) \\ g(Ma) &= \sum_{i=1}^{4} C_i \cdot Ma^i \end{aligned} \tag{17}$$

Table 4. Fitting coefficients of $g(Ma)$.

C_4	C_3	C_2	C_1
62.50	−933.84	3391.2	2024.4

$f(\Phi)$ reflects the influence of equivalence ratio on specific impulse. If the equivalence ratio is lower, fuel is more likely to burn completely due to more adequate oxygen; thus, larger thrust can be generated by a unit mass of fuel, and the specific impulse is larger as a result. Compared with the HL-20 aircraft model, whose impulse is independent from equivalence ratio [4,6], this model is closer to the actual situation.

$g(Ma)$ reflects the influence of Mach number on the specific impulse. When Mach number increases, specific impulse will decrease.

Figure 3 shows the comparison between the parametric thrust model and the original data at an angle of attack of 5°. It can be seen that the parametric model can accurately reflect the trend of thrust with Mach number and equivalence ratio: with the increase of Mach number, the thrust increases gradually; when the equivalence ratio increases, more fuel is burned and more thrust is generated.

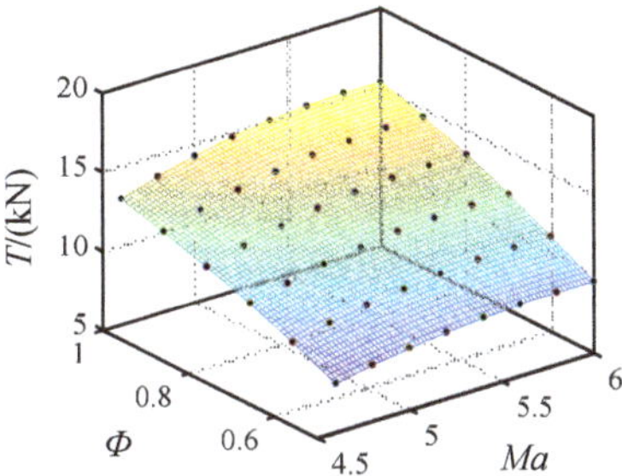

Figure 3. Comparison between the parametric model and original data.

2.4. Fuel Consumption Calculation

To evaluate the flight cost during steady-state cruise, the fuel consumption averaged by range is employed as an indicator. If the fuel consumption averaged by range is smaller, the range is larger using the same mass of fuel. The time derivative of range is [35]:

$$\frac{dr}{dt} = Ma \cdot c \cdot \cos \gamma \left(\frac{R_e}{R_e + h} \right) \tag{18}$$

Combined with Equations (10) and (14), the derivative of fuel consumption to range can be obtained as follows:

$$\frac{dm}{dr} = \frac{\Phi \cdot k \cdot \varphi_{inlet} \cdot \rho \cdot A}{\cos \gamma} \left(1 + \frac{h}{R_e} \right) \tag{19}$$

Due to the Mach number and altitude remaining unchanged during the steady-state cruise and the fuel consumption per second being rather small compared with the total aircraft mass, it is considered that all the parameters in Equation (19) remain basically unchanged [11]; thus, the fuel consumption averaged by range can be calculated by Equation (19) directly.

So far, the parametric aircraft model is established. Given the angle of attack, Mach number, and equivalence ratio, the aerodynamic force and thrust can be calculated, then the cruise trajectory can be obtained by the dynamic equation. Finally, the fuel consumption averaged by the range can be solved.

In the dynamic equation and parametric aircraft model, the relevant aircraft parameters include lift coefficient, drag coefficient, inlet area, specific impulse, aircraft mass, and reference area. In this paper, these six aircraft parameters are regarded as the parameter group that is mainly focused on.

3. Solution of Optimal Cruise Point

Previous studies have found that, when the aircraft is cruising at different Mach numbers and altitudes, its fuel consumption averaged by the range is different, and there is a certain value of altitude and Mach number to minimize the fuel consumption. According to the aircraft model in this paper, the steady-state cruise characteristics at different Mach numbers and altitudes are analyzed.

3.1. Steady-State Cruise Parameter Solution

In steady-state cruise, the elevation angle of trajectory is constant at 0, and the altitude and Mach number remain unchanged; thus, Equation (4) is simplified into:

$$\begin{cases} \frac{dMa}{dt} = \frac{T \cos \alpha - D - mg \sin \gamma}{m \cdot c} = 0 \\ \frac{d\gamma}{dt} = \frac{T \sin \alpha + L}{m \cdot Ma \cdot c} + \cos \gamma \left(\frac{Ma \cdot c}{R_e + h} - \frac{g}{Ma \cdot c} \right) = 0 \end{cases} \tag{20}$$

Lift and drag are related to altitude, Mach number, and angle of attack, while thrust is related to altitude, Mach number, angle of attack, and equivalence ratio. In short, there are two equations with four unknowns. From Equation (20), it can be obtained that

$$\begin{cases} T\cos\alpha - D - mg\sin\gamma = 0 \\ T\sin\alpha + L + \cos\gamma \cdot \dfrac{m\cdot(Ma\cdot c)^2}{R_e + h} - mg\cos\gamma = 0 \end{cases} \tag{21}$$

If the upper equation in Equation (21) is multiplied by $\tan\gamma$, T can be eliminated:

$$D\tan\alpha + L - mg + m\frac{(Ma\cdot c)^2}{(R_e + h)} = 0 \tag{22}$$

Then, substitute Equation (5) into Equation (22) and it can be obtained that

$$C_D \cdot \frac{1}{2}\rho(Ma\cdot c)^2 \cdot S \cdot \tan\alpha + C_L \cdot \frac{1}{2}\rho(Ma\cdot c)^2 \cdot S - mg + m\frac{(Ma\cdot c)^2}{(R_e + h)} = 0 \tag{23}$$

If the value of h and Ma are given, ρ and c can be obtained. Due to the fact that C_L and C_D are related to Ma and α, Equation (23) is an equation only related to α, which can be solved by the method of bisection. After the value of α is obtained, there is only one unknown left. Then, the equivalent ratio can be solved by Equation (21) so as to solve the fuel consumption.

The fuel consumption averaged by the range is solved in the range of Mach 4 to 7 and 20 km to 29 km by the traversing method at a Mach number interval of 0.1 and an altitude interval of 200 m. The contour of fuel consumption at different altitudes and Mach numbers is displayed in Figure 4. Since the equivalence ratio ranges from 0.45 to 1, there are boundaries in the contour caused by the constraints of the equivalence ratio, which means there is no solution of Equation (21) and steady-state cruise cannot be achieved in the region outside the boundaries. The distribution of fuel consumption is similar to a basin. If the Mach number or altitude is too high or too low, the fuel consumption will increase up to 1.2 kg/km, while the minimum is only about 0.4 kg/km. The group of the Mach number and altitude with the minimum fuel consumption is called the optimal steady-state cruise point and is denoted by (Ma_{opt}, h_{opt}), which is in the range of Mach 4.5 to 5.5 and 24 km to 26 km.

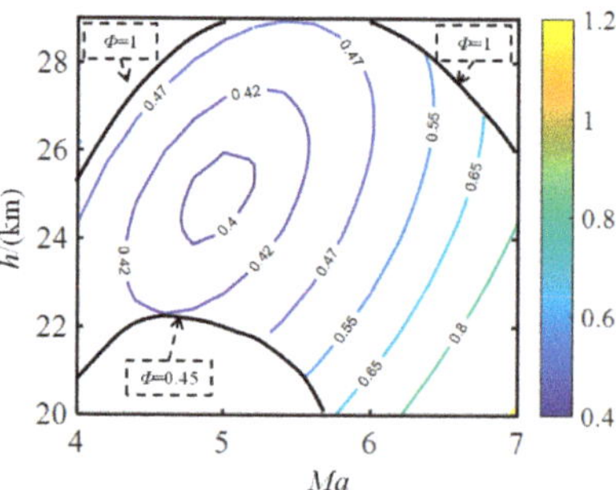

Figure 4. Contour of fuel consumption averaged by range.

3.2. Fast Solution of Optimal Cruise Point

Although a rough position of the optimal cruise point can be obtained by the traversing method, it is time-consuming and accuracy is limited. In order to determine the position of the optimal cruise quickly and accurately, the optimization algorithm is employed.

According to the analysis, there are actually only two variables when solving steady-state cruise. Given the cruise Mach number and altitude, the other trajectory parameters can be calculated. Therefore, the solution of the optimal cruise point can be regarded as an optimization problem with two variables in which the fuel consumption is the optimization objective and the cruise Mach number and altitude are the optimization variables. Therefore, the mathematical expression of the optimization problem is:

$$\begin{cases} \min J = \dfrac{\Phi \cdot k \cdot \varphi_{inlet} \cdot \rho \cdot A}{\cos \gamma}\left(1 + \dfrac{h}{R_e}\right) \\[2mm] s.t \begin{cases} 22 \leq h \leq 28 \\[1mm] 4 \leq Ma \leq 7 \\[1mm] \dfrac{T\cos\alpha - D - mg\sin\gamma}{m \cdot c} = 0 \\[2mm] \dfrac{T\sin\alpha + L}{mMa \cdot c} + \cos\gamma\left(\dfrac{Ma \cdot c}{R_e + h} - \dfrac{g}{Ma \cdot c}\right) = 0 \end{cases} \end{cases} \tag{24}$$

The particle swarm optimization (PSO) algorithm is used to solve the optimization problem. The optimization algorithm process is displayed in Figure 5, and the details are as follows:

Step 1: randomly generate an initial particle swarm, and each particle is two-dimensional and represents a group of Mach number and altitude;
Step 2: calculate the steady-state cruise trajectory parameters at the Mach number and altitude represented by a particle;
Step 3: calculate the fuel consumption averaged by range at the altitude and Mach number represented by a particle as the fitness function value;
Step 4: update the velocity and position of particles according to the fitness function value to obtain a new iteration of particle swarm;
Step 5: repeat Steps 2 to 5 until the termination conditions are satisfied.

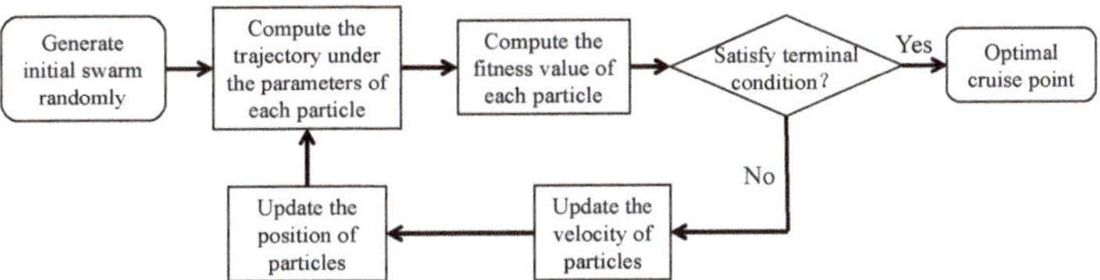

Figure 5. Optimization process.

Based on the optimization process, the optimal cruise point can be obtained. The size of the particle swarm is 10, and the maximum number of iterations is 30. Figure 6 shows the change of the fitness function value. The algorithm converges after 10 iterations. Finally, the altitude of the optimal cruise point is 24.95 km, and the Mach number is 4.95. Figure 7 displays the position of the optimal cruise point in the fuel consumption contour. The angle of attack corresponding to the optimal cruise point is 9.47°, and the equivalence ratio is 0.556. The minimal fuel consumption averaged by the range is 0.396 kg/km.

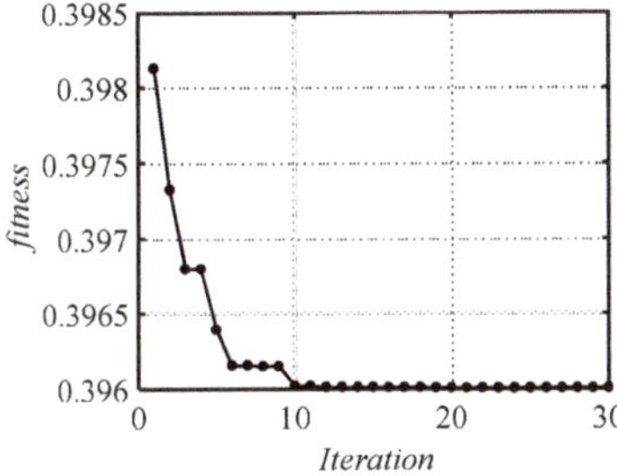

Figure 6. The change of fitness function value during optimization.

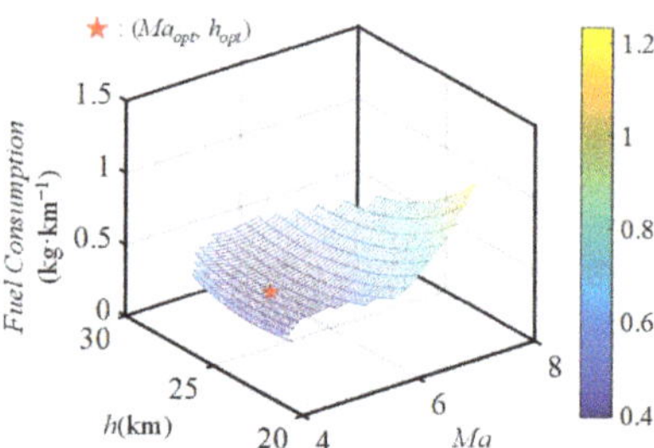

Figure 7. The location of optimal cruise point in the contour of fuel consumption.

Therefore, if the aircraft parameters are determined, the position of the optimal cruise point can be obtained quickly and accurately by this PSO algorithm. On this basis, the aircraft parameters can be adjusted and corresponding optimal cruise points can be obtained, then the influence of the aircraft parameters on the optimal cruise point can be explored.

4. Analysis and Modeling of Aircraft Parameters

According to previous analysis, for a determined parametric aircraft model, there is a corresponding optimal steady-state cruise point, and the position of the optimal cruise point is determined by the aircraft parameters. How the aircraft parameters affect the position of the optimal cruise point is also a problem worthy of exploration, which has guiding significance for the design of aircraft.

4.1. Analysis of Single Aircraft Parameter

The aircraft parameters group studied in this paper includes lift coefficient, drag coefficient, inlet area, specific impulse, aircraft mass, and reference area. These six parameters can be divided into three aspects, as shown in Figure 8: aerodynamic shape (lift coefficient and drag coefficient), propulsion system (inlet area and specific impulse), and structural design (aircraft mass and reference area).

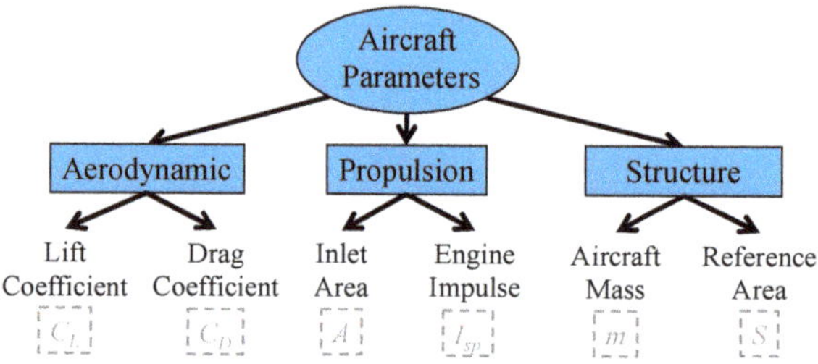

Figure 8. Considered parameters in this paper.

The values of these six aircraft parameters are adjusted to explore their influence on the optimal cruise point denoted by (Ma_{opt}, h_{opt}). The ratio between the value after adjustment and the original value is denoted by X_L, X_D, X_A, X_I, X_m, and X_S, respectively.

Within the range of 0.8 to 1.5, the influence of a single factor in (X_L, X_D, X_A, X_I, X_m, X_S) is investigated. Only one variable is changed at a time, and the rest remain at 1. After the aircraft parameters are adjusted, the corresponding optimal cruise points are obtained by the PSO algorithm in Figure 5.

Figure 9 shows the influence of different factors. From Figure 9a, Ma_{opt} has a positive correlation with X_A and X_I, and the effects of X_A and X_I on Ma_{opt} are very similar, which indicates that, if the magnitude of thrust determined by X_A and X_I is larger, Ma_{opt} will increase as well. Further, Ma_{opt} has a negative correlation with X_D and X_S, and the effects of X_D and X_S on Ma_{opt} are also similar, which indicates that, if the magnitude of drag determined by X_D and X_S is larger, Ma_{opt} is lower. Differently, X_L and X_m have little effect on Ma_{opt}.

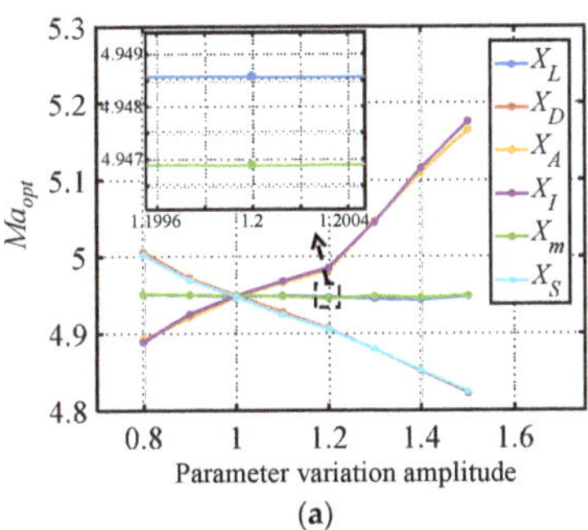
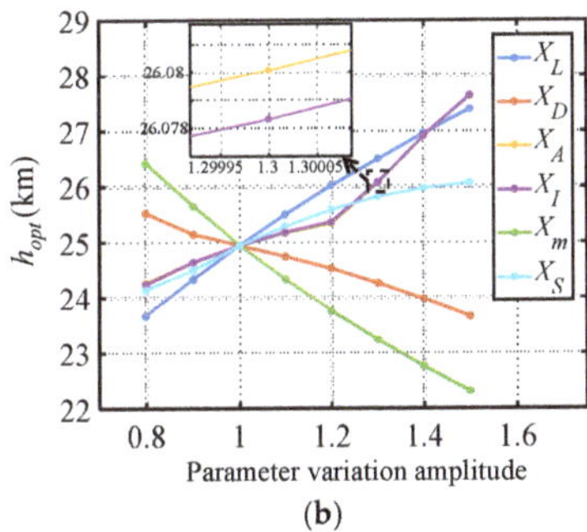

Figure 9. Influence of different factors on the optimal cruise point: (**a**) Influence on Ma_{opt}; (**b**) Influence on h_{opt}.

It can be seen from Figure 9b that h_{opt} is sensitive to all the factors. In detail, h_{opt} is positively correlated with X_L, X_A, X_I, and X_S, and the effects of X_A and X_I on h_{opt} are very similar, which indicates that the magnitude of thrust determined by X_A and X_I also has a great impact on h_{opt}, while h_{opt} has a negative correlation with X_D and X_m.

A multiple regression analysis is employed to quantify the influence of the six factors on the optimal cruise point, and the regression coefficient is normalized as an index of sensitivity. Figure 10 shows the sensitivity percentages of the different factors, where a negative sign indicates a negative correlation between the factor and the dependent variable. It can be seen that X_A and X_I have the largest impact on the optimal cruise Mach number, accounting for nearly 30%, followed by X_D and X_S, which are negatively correlated, while the sensitivity of X_L and X_m is almost 0; X_A, X_I and X_m have the greatest influence on the optimal cruise altitude, accounting for about 20%, where the influence of X_m is negative.

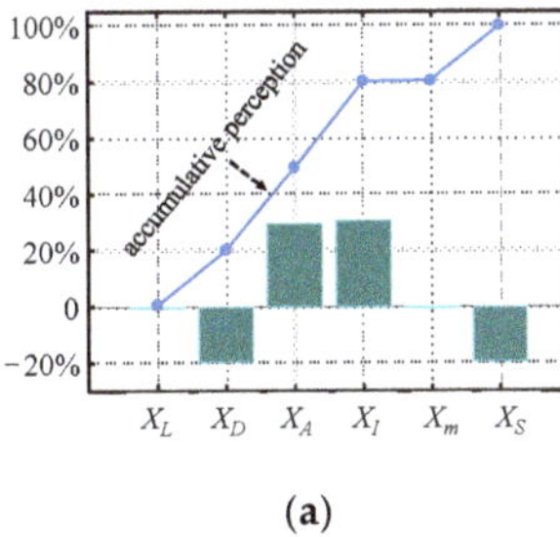
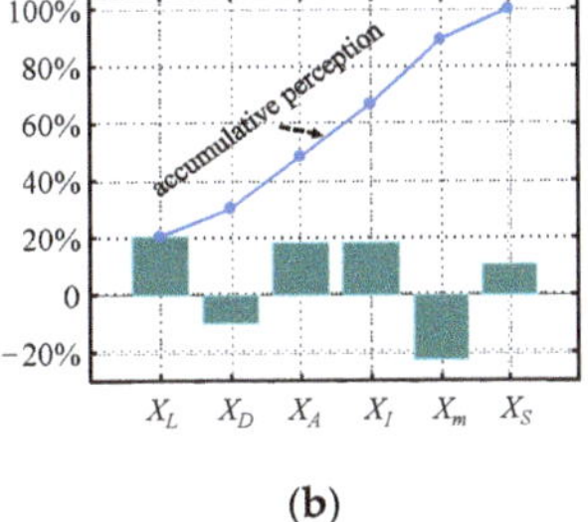

Figure 10. Sensitivity of different factors on the optimal cruise point: (**a**) Sensitivity on Ma_{opt}; (**b**) Sensitivity on h_{opt}.

Based on the conclusions above, the inlet area, specific impulse, and drag coefficient have the largest impact on the optimal steady-state cruise Mach number. If the optimal cruise Mach number needs to be enhanced, the inlet area and specific impulse should be larger or the drag of aircraft should be reduced. Due to the inlet area, the specific impulse and aircraft mass have the largest impact on the optimal cruise altitude. To improve the optimal cruise altitude, the most effective way is to increase the inlet area and specific impulse, or reduce the weight of the aircraft.

Figure 11 shows the influence of the six factors on fuel consumption averaged by the range. It can be seen that X_D, X_m, and X_S are positively correlated with fuel consumption, while X_L, X_A, and X_I are negatively correlated with it. Therefore, in order to reduce the fuel consumption, the lift coefficient and specific impulse as well as the inlet area should be increased, where the effect of increasing specific impulse is the best; in addition, the mass and drag coefficient, as well as the reference area of the aircraft, should be reduced, where the effect of drag coefficient reduction is the best.

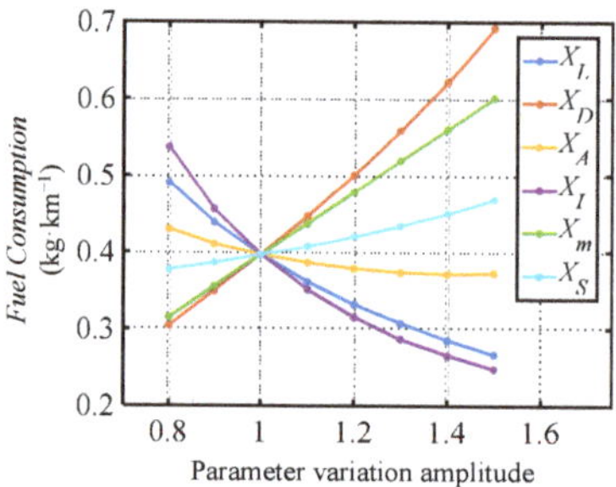

Figure 11. Influence of different influence factors on the fuel consumption.

However, compared with the altitude and fuel consumption of the optimal cruise point, the change range of the Mach number is the smallest, which means that it is difficult to greatly change the optimal cruise Mach number by adjusting the aircraft parameters in the current range.

4.2. Modeling of Aircraft Parameters

Through the analysis of a single factor, the influence of different aircraft parameters on the optimal cruise point is clarified. However, due to the complex interaction among these parameters, the influence of a single factor has limited ability to guide aircraft design, and it is necessary to consider simultaneous changes of different aircraft parameters. Therefore, it is of significance to establish a mathematical model that can comprehensively describe the relationship between the aircraft parameters and optimal cruise point.

A neural network (NN) is a complex network system that simulates the human brain. It consists of a large number of simple neurons connected with each other [36–38]. A feed forward neural network (FFNN) is a kind of neural network. In recent years, it has been widely used in data predicting and so on [39]. In this paper, due to there being six parameters, it is difficult to model the influence law by polynomial fit explicitly. Therefore, FFNN is employed to establish a mathematical model of (Ma_{opt}, h_{opt}) and $(X_L, X_D, X_A, X_I, X_m, X_S)$; the model has six inputs and two outputs.

Firstly, 500 sample points are generated for the six inputs from the Optimal Latin Hypercube Distribution [40] in a range of 0.8 to 1.5, and the values of (Ma_{opt}, h_{opt}) for the 500 samples are obtained by the PSO for the optimal cruise point in Figure 5. Furthermore, 400 of the samples are used as the training set, 75 as the validation set, and the other 25 as the test set. The logsig function is employed in the hidden layer, and the output layer transfer function is Purelin Linear. The L-M method is employed as the learning method. The structure of the FFNN is displayed in Figure 12.

The regression results after training are illustrated in Figure 13. It can be seen that the target value and output results are basically on the same line, and the value of the regression coefficient is close to 1, which shows a good training effect.

The model estimated values and accurate values of the 25 points in the test set are compared in Figure 14. It can be seen that the estimated values are within the deviation range of 3% of the accurate value, indicating that the trained neural network model can accurately describe the relationship between (Ma_{opt}, h_{opt}) and $(X_L, X_D, X_A, X_I, X_m, X_S)$.

Based on this model, given the value of the six factors, the values of (Ma_{opt}, h_{opt}) can be quickly obtained.

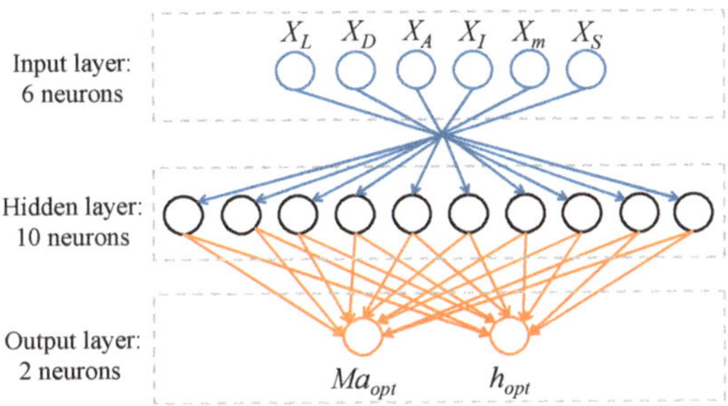

Figure 12. The structure of FFNN.

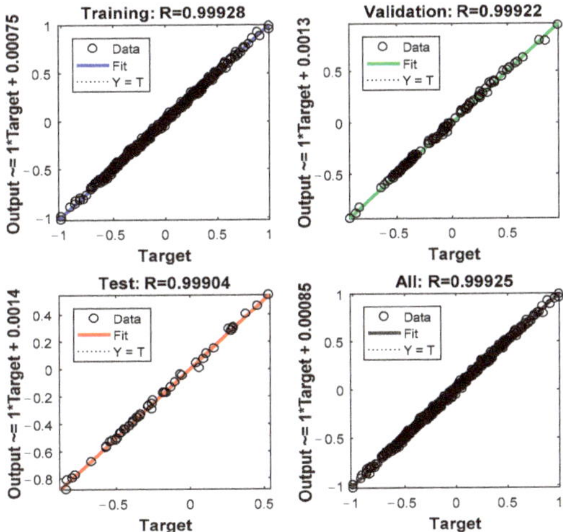

Figure 13. Regression results of FFNN after training.

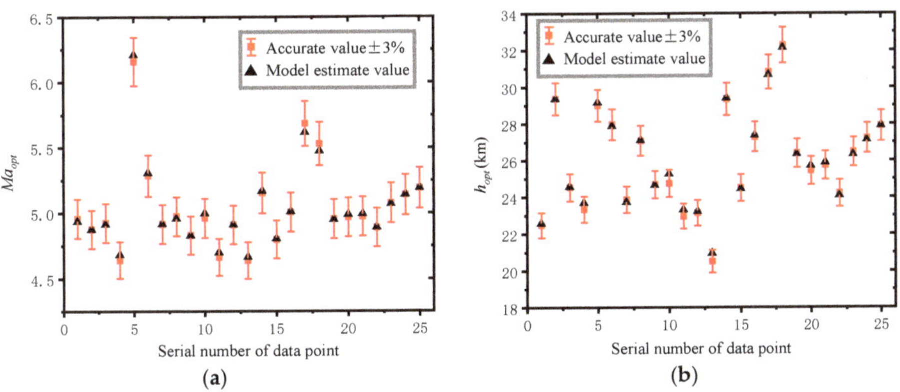

(a)

(b)

Figure 14. Comparison between estimated value and accurate value: (**a**) Estimated value and accurate value of Ma_{opt}; (**b**) Estimated value and accurate value of h_{opt}.

5. Aircraft Parameter Optimization Method

After the model is obtained, given the value of $(X_L, X_D, X_A, X_I, X_m, X_S)$, (Ma_{opt}, h_{opt}) can be obtained quickly. Therefore, if various parameters change simultaneously, the reaction of the optimal cruise point can be studied. By this way, if the subjective desired cruise point does not coincide with the actual optimal cruise point, the optimal cruise point can be adjusted to the target position by optimizing the aircraft parameters.

However, due to the interaction between the aircraft parameters, the coupling relationship between the aircraft parameters needs to be considered when optimization is carried out.

Since there is a complex coupling relationship between the aircraft parameters, take the following constraints as examples to illustrate the optimization method of the aircraft parameters based on the neural network model:

(1) The maximum increment of the lift coefficient is 5%, and the drag coefficient will also increase at the same time, and the minimal increment is 1/2 of that of the lift coefficient;

(2) The maximum increment of the inlet area is 30%; meanwhile, the aircraft mass and reference area will also increase, and the minimal increments are 1/3 and 1/4 of that of the inlet area, respectively;

(3) The maximum increment of the specific impulse is 10%.

These constraints can be adjusted according to the actual situation. Now that it is difficult to adjust the optimal cruise Mach number greatly, the optimal cruise altitude is mainly focused. The target of the optimal cruise point is set at $(Ma5, 26\ km)$.

However, if the aircraft parameters can be changed at the same time, to adjust the optimal cruise altitude to the desired value, there is more than one plan since six relevant aircraft parameters are considered. The fuel consumption after aircraft parameters optimization is different by various adjustment plans. Therefore, in this paper, the plan with the minimum fuel consumption after adjustment is explored. Since there may be more than one solution, under the condition that the optimal cruise point after adjustment is located at $(Ma5, 26\ km)$, reducing the fuel consumption is also an objective. The condition that the optimal cruise point is located at $(Ma5, 26\ km)$ is reflected in the fitness function in the form of the penalty function, so the fitness function is written as:

$$fitness = F_h + F_M + \frac{\widetilde{T}}{g\widetilde{I_{sp}}\widetilde{Ma} \cdot c \cdot \cos\gamma}\left(1 + \frac{\widetilde{h}}{R_e}\right) \tag{25}$$

where "~" indicates that the parameter after adjustment, and the penalty functions F_h and F_M are regarding altitude and Mach number, respectively.

$$F_h = \begin{cases} \lambda_1 \cdot \frac{|\widetilde{h}_{opt}-26|}{26}, & if\ \frac{|\widetilde{h}_{opt}-26|}{26} > \varepsilon \\ 0, & else \end{cases}$$
$$F_M = \begin{cases} \lambda_2 \cdot \frac{|\widetilde{Ma}_{opt}-5|}{5}, & if\ \frac{|\widetilde{Ma}_{opt}-5|}{5} > \varepsilon \\ 0, & else \end{cases} \tag{26}$$

λ_1 and λ_2 are both large numbers, and ε is a small tolerance.

Therefore, the aircraft parameter optimization problem can be expressed mathematically as:

$$\text{minimize } fitness = F_h + F_M + \frac{\tilde{T}}{g \tilde{I}_{sp} \widetilde{Ma} \cdot c \cdot \cos\gamma}\left(1 + \frac{\tilde{h}}{R_e}\right)$$

$$\text{subject to } \begin{cases} X_L \leq 1.05 \\ X_D \geq 1 + \frac{1}{2}(X_L - 1) \\ X_A \leq 1.3 \\ X_m \geq 1 + \frac{1}{3}(X_A - 1) \\ X_S \geq 1 + \frac{1}{4}(X_A - 1) \\ X_I \leq 1.1 \end{cases} \tag{27}$$

The particle swarm optimization algorithm is employed, whose process is outlined in Figure 5. Each particle in the algorithm represents a group of $(X_L, X_D, X_A, X_I, X_m, X_S)$, and the objective is to minimize the fitness of the particles. The optimal solution should be a group of $(X_L, X_D, X_A, X_I, X_m, X_S)$, with the lowest fitness computed by (25).

The results are displayed in Table 5. The lift coefficient needs to increase by 5%, while the drag coefficient will increase by 2.5% as a result; the inlet area should increase by 25.93%, while the mass will increase by 8.64%, and the reference area will increase by 12.91%; the specific impulse needs to increase by 10%. It can be seen that the variation of the lift coefficient and impulse reach their boundaries, while the variation of the inlet area and reference area are within the feasible range. If the constraints are different, the optimization results will change as well. After the aircraft parameters are adjusted, the optimal steady-state cruise point estimated by the neural network model is located at ($Ma4.99, 25.95$ km).

Table 5. Optimized results of aircraft parameters.

X_D	X_I	X_L	X_m	X_S	X_A
1.025	1.1	1.05	1.0864	1.1291	1.2593

Under the optimized aircraft parameters, the position of the accurate optimal cruise point and contour of fuel consumption are displayed in Figure 15. It can be seen that the actual optimal cruise point after adjustment is located at ($Ma4.99, 25.92$ km), which basically meets the goal of adjusting the optimal cruise point from ($Ma4.95, 24.95$ km) to ($Ma5, 26$ km). In addition, the adjustment adopts a plan with the lowest fuel consumption at 0.3642 kg/km with the constraints satisfied. Therefore, the method of optimizing the aircraft parameters based on the neural network model has been demonstrated to be effective.

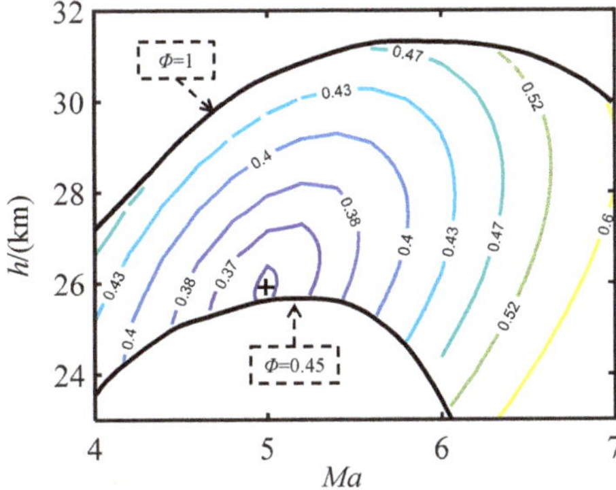

Figure 15. Optimal point and contour of fuel consumption after adjustment.

So far, not only the optimal cruise point is adjusted to the target position but also the fuel consumption after adjustment is the lowest. However, it is difficult to greatly adjust the optimal cruise Mach number under the current constraints. If the technical bottleneck of the ramjet can be broken through and an engine with a larger size and significantly increased specific impulse can be obtained, it will be of great benefit to adjust the optimal cruise Mach number in a large range.

6. Conclusions

There is an optimal cruise point for hypersonic vehicles when performing steady-state cruise, and, at this point, the fuel consumption is the lowest. The position of the optimal cruise point is closely related to the aircraft parameters. In this paper, to clarify the influence of the aircraft parameters on the optimal cruise point, the optimal cruise point under a basic parametric aircraft model was firstly solved by an optimization algorithm, and then the influence of different parameters was investigated with modeling by a neural network. In order to make the aircraft cruise at the desired Mach number and altitude with the lowest fuel consumption, an aircraft parameter optimization method whose aim was to adjust the optimal cruise point to a given position was explored with numerical vindication. The main conclusions of this paper are as follows:

(1) The influence of aircraft parameters on the optimal cruise point is clarified. The optimal cruise Mach number is mainly related to the specific impulse and inlet area, and the optimal cruise altitude is mainly related to the specific impulse, inlet area, and aircraft mass.
(2) The neural network model established in this paper can accurately describe the influence of the aircraft parameters on the position of the optimal cruise point.
(3) The aircraft parameter optimization method is effective to adjust the optimal cruise point to a desired position, and the adjustment plan is also optimal in minimizing the fuel consumption, which provides a new perspective for the design of aircraft.

The parameter analysis and optimization method can be applied in other aircraft models, and the constraints of the aircraft parameters can be changed according to the actual situation as well, which can reveal how the aircraft parameters should be adjusted and thus provide guidance in the design of aircraft.

Author Contributions: Conceptualization, H.L. and Y.W.; Investigation, Y.W.; Methodology, H.L. and Y.Z.; Software, Y.Z. and S.D.; Visualization, S.D.; Writing—original draft, H.L.; Writing—review & editing, S.X. All authors have read and agreed to the published version of the manuscript.

Funding: This research was funded by Postgraduate Scientific Research Innovation Project of Hunan Province, grant number CX20200084.

Institutional Review Board Statement: Not applicable.

Informed Consent Statement: Not applicable.

Data Availability Statement: The data presented in this study are available on request from the corresponding author.

Acknowledgments: The authors want to express their thanks to Tao Tang.

Conflicts of Interest: The authors declare no conflict of interest.

References

1. Fan, Y.; Lu, F.; Zhu, W.; Bai, G.; Yan, L. A hybrid model algorithm for hypersonic glide vehicle maneuver tracking based on the aerodynamic model. *Appl. Sci.* **2017**, *7*, 159. [CrossRef]
2. Li, Z.; Shi, S. Adaptive loss fault tolerance control of unmanned hypersonic aircraft with elasticity. *Aerospace* **2021**, *8*, 176. [CrossRef]
3. Fan, Y.; Zhu, W.; Bai, G. A Cost-effective tracking algorithm for hypersonic glide vehicle maneuver based on modified aerodynamic model. *Appl. Sci.* **2016**, *6*, 312. [CrossRef]
4. Li, H.; Wang, Y.; Zhou, Y.; Xu, S.; An, K.; Fan, X. Optimization and analysis for hypersonic steady-state cruise trajectory. In Proceedings of the 33rd Chinese Control and Decision Conference, Kunming, China, 22–24 May 2021; pp. 7555–7560.

5. Wang, K.; Zhang, B.; Hou, Y. Multiobjective optimization of steady-state cruise trajectory for a hypersonic vehicle. In Proceedings of the 2017 3rd IEEE International Conference on Control Science and Systems Engineering, Beijing, China, 17–19 August 2017; pp. 130–135. [CrossRef]
6. Chuang, C.; Morimoto, H. Periodic optimal cruise for a hypersonic vehicle with constraints. *J. Guid. Control Dyn.* **1997**, *34*, 165–171. [CrossRef]
7. Wang, W.; Hou, Z.; Shan, S.; Chen, L. Optimal periodic control of hypersonic cruise vehicle: Trajectory features. *IEEE Access* **2019**, *7*, 3406–3421. [CrossRef]
8. Chen, R.; Williamson, W.; Speyer, J.; Youssef, H. Optimization and implementation of periodic cruise for a hypersonic vehicle. *J. Guid. Control Dyn.* **2006**, *29*, 1032–1040. [CrossRef]
9. Kang, B.; Spencer, D.; Tang, S.; Jordan, D. Optimal periodic cruise trajectories via a two-level optimization method. *J. Spacecr. Rockets.* **2015**, *47*, 597–613. [CrossRef]
10. Kang, B.; Tang, S. Trajectory design and optimization of powered common aero vehicles. *Acta Aeronaut. Astronaut. Sin.* **2009**, *30*, 738–743. [CrossRef]
11. Liu, X.; Wang, Y.; Liu, L.; Huang, J. Optimization on steady-state cruise for a hypersonic vehicle. In Proceedings of the 33rd Chinese Control Conference, Nanjing, China, 28–30 July 2014; pp. 8935–8940. [CrossRef]
12. Gao, H.; Chen, Z.; Sun, M.; Chen, Z. General periodic cruise guidance optimization for hypersonic vehicles. *Appl. Sci.* **2020**, *10*, 2898. [CrossRef]
13. Li, H.; Wang, Y.; Zhou, Y.; Xu, S.; Su, D. A two-level optimization method for hypersonic periodic cruise trajectory. *Int. J. Aerosp. Eng.* **2021**, *2021*, 9975007. [CrossRef]
14. Zhang, H.; Gao, F.; Li, X.; Zhang, H. Numerical study on the effects of the equivalence ratio on the performance of supersonic combustor. *J. Solid Rocket Technol.* **2015**, *4*, 487–491. [CrossRef]
15. Lewis, M. Significance of fuel selection for hypersonic vehicle range. *J. Propul. Power.* **2001**, *17*, 1214–1221. [CrossRef]
16. Luo, J.; Li, C.; Xu, J. Inspiration of hypersonic vehicle with airframe/propulsion integrated design. *Acta Aeronaut. Astronaut. Sin.* **2015**, *36*, 39–48. [CrossRef]
17. Luo, J.; Xu, M.; Liu, J. Research on lift and drag characteristic for the integrated configuration of hypersonic vehicle. *J. Astronaut.* **2007**, *28*, 1478–1481. [CrossRef]
18. Zhang, Z.; Gao, Y.; Baoyin, H. Study on relationship between periodic cruise and configuration parameters of hypersonic vehicle. *Missile Space Veh.* **2009**, *4*, 4–7. [CrossRef]
19. Chen, B.; Xu, G. Comparison of aerodynamic characteristic of near space high lift-drag ratio configuration. *Chin. Q. Mech.* **2014**, *35*, 464–472. [CrossRef]
20. Isaienko, V.; Kharchenko, V.; Matiychyk, M.; Lukmanova, I. Analysis of layout and justification of design parameters of a demonstration aircraft based on solar cells. *E3S Web Conf.* **2020**, *164*, 13007. [CrossRef]
21. Li, G.; Wang, J.; Wang, Y. Parametric research of the buoyancy-lifting aerial vehicle with the low pressure energy storage method. *Acta Aeronaut. Astronaut. Sin.* **2021**, *42*, 376–388. [CrossRef]
22. Queipo, N.; Haftka, R.; Shyy, W.; Goel, T.; Vaidyanathan, R.; Kevin Tucker, P. Surrogate-based analysis and optimization. *Prog. Aerosp. Sci.* **2005**, *41*, 1–28. [CrossRef]
23. Zhang, F.; Gao, Y.; Yao, H.; Wang, Y.; Luo, K. Structural Parameter Sensitivity Analysis of an Aircraft Anti-Icing Cavity Based on Thermal Efficiency. *Int. J. Aerosp. Eng.* **2019**, *2019*, 7851260. [CrossRef]
24. Luo, S.; Luo, W.; Wang, Z. Analysis of the sensitivity of hypersonic cruise vehicle airframe/propulsion system integrated design parameters. *J. Natl. Univ. Def. Technol.* **2003**, *25*, 10–14. [CrossRef]
25. Deng, F.; Qiao, Z.; Fu, Q.; Chen, L.; Tian, S.; Zhang, D. Hypersonic vehicle shape optimal design based on overall performance. *J. Beijing Univ. Aeronaut. Astronaut.* **2016**, *42*, 2587–2595. [CrossRef]
26. Kim, S.; Yun, J.; Hwang, H. Sensitivity analysis and technology evaluation for a roadable personal air vehicle at the conceptual design stage. *Appl. Sci.* **2019**, *9*, 4121. [CrossRef]
27. Roy, A.G.; Peyada, N.K. Aircraft parameter estimation using Hybrid Neuro Fuzzy and Artificial Bee Colony optimization (HNFABC) algorithm. *Aerosp. Sci. Technol.* **2017**, *71*, 772–782. [CrossRef]
28. Forrester, A.; Keane, A. Recent advances in surrogate-based optimization. *Prog. Aerosp. Sci.* **2009**, *45*, 50–79. [CrossRef]
29. Lian, Y.; Oyama, A.; Liou, M. Progress in design optimization using evolutionary algorithms for aerodynamic problems. *Prog. Aerosp. Sci.* **2010**, *46*, 199–223. [CrossRef]
30. Pontani, M.; Conway, B. Particle Swarm Optimization Applied to Space Trajectories. *J. Guid. Control Dyn.* **2010**, *33*, 1429–1441. [CrossRef]
31. Kim, J.; Lee, J. Trajectory Optimization with Particle Swarm Optimization for Manipulator Motion Planning. *IEEE Trans. Industr. Inform.* **2015**, *11*, 620–631. [CrossRef]
32. Guo, Z.; Qu, Z.; Liu, J. Numerical determination of trim angle of attack for reentry vehicles using dynamic unstructured grids. *Acta Aerodyn. Sin.* **2004**, *22*, 255–258. [CrossRef]
33. Li, H.; Wang, Y.; Xu, S.; Zhou, Y.; Su, D. Trajectory optimization of hypersonic periodic cruise using an improved PSO algorithm. *Int. J. Aerosp. Eng.* **2021**, *2021*, 2526916. [CrossRef]
34. Xu, W.; Lu, Y.; Liu, Y.; Chen, B. Design and optimization of periodic cruise trajectory for hypersonic vehicle. *Flight Dyn.* **2016**, *34*, 33–36. [CrossRef]

35. Wang, Z.; Cheng, X.X.; Li, H. Hypersonic Skipping Trajectory Planning for High L/D Gliding Vehicles. In Proceedings of the 21st AIAA International Space Planes and Hypersonic Technologies Conference, Xiamen, China, 6–9 March 2017; pp. 1–7. [CrossRef]
36. Izadpanahkakhk, M.; Razavi, S.M.; Taghipour-Gorjikolaie, M.; Zahiri, S.H.; Uncini, A. Deep Region of Interest and Feature Extraction Models for Palmprint Verification Using Convolutional Neural Networks Transfer Learning. *Appl. Sci.* **2018**, *8*, 1210. [CrossRef]
37. Erskine, S.K.; Elleithy, K.M. Real-Time Detection of DoS Attacks in IEEE 802.11p Using Fog Computing for a Secure Intelligent Vehicular Network. *Electronics* **2019**, *8*, 776. [CrossRef]
38. Tao, J.; Sun, G. An artificial neural network approach for aerodynamic performance retention in airframe noise reduction design of a 3D swept wing model. *Chin. J. Aeronaut.* **2016**, *29*, 1213–1225. [CrossRef]
39. Franco, S.; Ah, C. Universal approximation using feedforward neural networks: A survey of some existing methods, and some new results. *Neural Netw.* **1998**, *11*, 15–37. [CrossRef]
40. Park, J. Optimal latin-hypercube designs for computer experiments. *J. Stat. Plan. Infer.* **1994**, *39*, 95–111. [CrossRef]

MDPI

St. Alban-Anlage 66

4052 Basel

Switzerland

Tel. +41 61 683 77 34

Fax +41 61 302 89 18

www.mdpi.com

Applied Sciences Editorial Office

E-mail: applsci@mdpi.com

www.mdpi.com/journal/applsci

www.ingramcontent.com/pod-product-compliance
Lightning Source LLC
LaVergne TN
LVHW070651170726
843515LV00004B/379